Agnelo Miguel

Agro-forestry systems for the recovery of degraded areas

Agnelo Miguel

Agro-forestry systems for the recovery of degraded areas

"The case of the villages of Calue and Cambongue"Huambo/Angola

ScienciaScripts

Imprint
Any brand names and product names mentioned in this book are subject to trademark, brand or patent protection and are trademarks or registered trademarks of their respective holders. The use of brand names, product names, common names, trade names, product descriptions etc. even without a particular marking in this work is in no way to be construed to mean that such names may be regarded as unrestricted in respect of trademark and brand protection legislation and could thus be used by anyone.

Cover image: www.ingimage.com

This book is a translation from the original published under ISBN 978-620-6-75810-5.

Publisher:
Sciencia Scripts
is a trademark of
Dodo Books Indian Ocean Ltd. and OmniScriptum S.R.L publishing group

120 High Road, East Finchley, London, N2 9ED, United Kingdom
Str. Armeneasca 28/1, office 1, Chisinau MD-2012, Republic of Moldova, Europe
Printed at: see last page
ISBN: 978-620-7-62061-6

Contents

INTRODUCTION

The evolution of agricultural production in Huambo province since the end of the war has followed the general trend in the country. In fact, large swathes of peasant families, whose agricultural production does not fully cover their food needs, are therefore looking for alternative sources of income outside of farming. In this sense, the phenomenon of pluriactivity plays an important role in their social reproduction; hence, almost everywhere in the villages there are cases of young people temporarily emigrating to big cities, mainly Luanda, where they find business opportunities and jobs, thus obtaining financial resources.

Soil degradation and declining fertility generate the need for new deforestation and have led to disorganised land occupation as a result of demographic growth and increased demand for land use. As a result, there has been a rural exodus, the concentration of income and land ownership for the survival of the household, as well as the persistence of cycles of poverty for small farmers in tropical countries. This has led to the need to look for new development models based on the sustainable use of natural resources, especially soil.

Farmers have a diversified system of self-consumption based on the intercropping of maize, beans, potatoes, vegetables and small livestock. They predominantly have areas of less than eight hectares and a low level of capitalisation. If they have surplus produce, they sell it on the local market. The families of these farmers have difficulty accessing pluriactivity due to the predominance of children in the family and or families with few members, the elderly and/or widows. Farm income and family income are very low.

In these circumstances, farmers are almost forced to clear new areas of forest and the farmers of Calue and Cambongue are an example of this situation. Agro-forestry systems are a good option for small rural producers and can reconcile food production with the conservation of natural resources and their maintenance.

SCIENTIFIC PROBLEM

The removal of vegetation causes serious environmental degradation problems and the soil is greatly affected as it is exposed to erosive agents, especially water erosion, which is considered one of the main causes of the degradation of natural resources.

GENERAL OBJECTIVE

To make a diagnosis of the agro-environmental situation in the villages of Calue, Cambongue, Lungongo and Cacaca, belonging to the municipality of Caala, and to analyse the possibility of these farmers adopting and adapting Agro-Forestry Systems (AFS) as a way of recovering the areas.

SPECIFIC OBJECTIVES

1. To diagnose the practices used in the production units of peasant families through surveys.
2. Evaluate by analysing the data obtained through surveys relating to:

- Type of plough
- Cultural practices used
- Soil fertility maintenance methods
- Adherence to Agroforestry Systems
- The main species to include in these Agro-Forestry Systems.

3. Proposing Agro-Forestry Systems for the recovery of degraded areas

HYPOTHESIS

The use of Agro-Forestry Systems can help restore degraded areas, improve soil fertility, diversify crops and reduce the effects of climate change.

CHAPTER 1

1. LITERATURE REVIEW

1.1. Concept and implications of an agricultural system

According to Mazoyer (1987) an agrarian system is first and foremost a historically constituted way of exploiting the environment, a system of forms of production, a technical system adapted to the bio-climatic conditions of a given area, which responds to the social conditions and needs of the time. A way of exploiting the environment that is the specific product of agricultural labour, using an appropriate combination of inert means of production and living means to exploit and reproduce a cultivated environment, resulting from the successive transformations historically undergone by the natural environment. An agricultural system can therefore be defined as a combination of the following essential variables:

- **The cultivated environment** - the original environment and its historical transformations;
- **Product instruments** - tools, machines, biological materials (cultivated plants, domestic animals, etc.) - and **the social labour force** (physical and intellectual) that uses them;
- **The** resulting **"artificialisation" of the environment** (the reproduction and exploitation of the cultivated ecosystem);
- **The social division of labour** between agriculture, handicrafts and industry, which allows for the reproduction of the instruments of labour and, consequently..;
- **Agricultural surpluses,** which, in addition to the needs of producers, make it possible to satisfy the needs of other social groups;
- **The relations of exchange** between the associated branches, the relations of ownership and the relations of force that regulate the distribution of the products of labour, product goods and consumer goods, and the relations of exchange between the systems (competition);
- In short, it's the **set of ideas and institutes** that make it possible to assume social reproduction: product, relation of product and exchange, distribution of the product, etc.

It is thanks to this concept that we can grasp and characterise the changes in the state of agriculture and the qualitative changes in the variables and their relationships, and develop a theory that allows us to distinguish, order and understand the great moments in the historical evolution and geographical differentiation of agrarian systems.
In this context, an Agro-Forestry System is defined as a system differentiated by having an arboreal or woody component, which plays a fundamental role in its structure and function (Medrado, 2000). Agro-forestry systems (AFSs) have the attributes of any system: **boundaries, components, interactions, inputs and outputs, hierarchical relationships and their own dynamics**. According to OTS/CATIE (1986), the boundaries are the physical borders of the system; the components are the physical, biological and socio-economic elements; the inputs and outputs are the material and energy that is transferred between different systems; the interactions are the relationships or energy and material that are exchanged between the components of the system; the hierarchy is the position and interrelationships with other systems.

1.2. Agroforestry Systems: Concept and Evolution

Tracing a historical line, the first origins of what is now known as the Agro-Forestry System (AFS) can be found in the past - in a wide geographical range. Almeida (2001) provides a summary of this geographical and anthropological breadth, when he mentions annual crops in deciduous forests in medieval Europe; the simulation of forest conduits through the planting of trees with different growth habits by Amerindian peoples; the complex and sophisticated itinerant cultivation in Asia, composing systems with rice and native trees, the latter being indispensable; and when he talks about subsistence crops with tree species in Nigeria and Zambia (Africa).
Much of the knowledge and foundations of agroforestry systems are the fruit of people's empiricism and are not systematised or explained (Peneireiro, 2006). In traditional systems, trees were kept in the system as a support and were also used to produce food. Technological innovations have simplified agricultural production systems and as a result SAFs have become less intense (Christina, 2002). According to Altieri (2002) "the potential of SAFs is recognised particularly by small farmers in poor and marginal areas of tropical and subtropical regions", including Asia, Africa and Latin America.
Altieri (2002) analysed SAFs and concluded that "whatever the concept, the essence is the use of agricultural and forestry elements in the same area in sustainable production systems" based on the integrated use of the land, suitable for marginal areas and the low use of production factors". Based on this concept, the author cla ssified Agro-Forestry Systems, dividing them into four types according to the criteria of composition and arrangement of the components on a functional, productive and conservationist basis:

- Agroforestry - the use of land for the simultaneous or sequential production of annual crops and forests;
- Agroforestry systems - a system in which the land is managed for the simultaneous production of agricultural and forestry crops and the rearing of domestic animals;
- Silvopastoral systems - management systems where forests are used to raise domestic animals as well as to produce timber, food and fodder;
- Multi-use forest production systems - a system in which trees are regenerated or managed to produce not only wood but also leaves and/or fruit suitable for food and/or fodder.

According to Peneireiro (2006) there are Agro-Forestry Systems based on different paradigms. According to Varejao (2009), there are Agro-Forestry Systems that only feature simple consortia between species, following the same paradigm of competition as monocultures; Agro-Forestry backyards and more complex Agro-Forestry formations, based on the dynamics of the forest itself. Vivan (1998) distinguishes between two lines of thought on Agro-Forestry Systems: conventional SAFs and regenerative analogue SAFs. The author distinguishes between these two systems by arguing that conventional SAFs are monotonous systems of monocultures in strata, while regenerative SAFs are more dynamic, complex and diverse systems. Agro-forestry systems based on forest ecology - defined as successional or regenerative analogues - seek to establish a dynamic of forms, nutrient cycling and dynamic equilibrium analogous to the original vegetation of the ecosystem (Vivan, 1998) and can be defined as complex systems, as they are made up of many components and these components are capable of exercising great autonomy when they can perform their function in the system well, even if other components are not working well. Such systems are highly diverse and suggest a high density of plants, which makes them adapted to tropical regions in multi-strata, thus optimising forest space (Campelo *et al*, 2006). The adaptation of SAFs in tropical regions is due to the climatic characteristics (temperature, radiation and precipitation) of the tropics and subtropics, which are suitable for the development of forests and biodiverse Agroforestry Systems. According to Bolfe *et al* (2003), SAFs are the interface between agriculture and forests, and as well as having ecologically correct characteristics, they bring human beings closer to the forest environment.

1.2.1 Advantages of Agroforestry Systems

Many environmental advantages are attributed to SAFs due to the similarity of their mechanisms with those of nature (Golley, 1983), from the rescue and maintenance of soil fertilisation to the natural control of pests and the use of their own components as inputs. In addition, socio-economic advantages are also considered, such as diversified production, which can be directed towards both domestic consumption and commercialisation; the possibility of commercialisation throughout the year due to staggered harvests and a reduction in risks due to products being affected differently by adverse weather conditions.

Because successional SAFs are based on the dynamics of the natural succession of species to leverage and compose them in the different phases (Vivan, 1998), this form of cultivation offers elements for the diversification of agricultural systems which, once established and well managed, will be governed by nature's own self-regulation. With regard to the soil, Howard (2007) describes nature's own soil management methods, exemplifying them through the dynamics of woods and forests and stating that these methods are the starting point for maintaining soil fertility, which for the author "is the first condition of any permanent cultivation in agriculture" and is the example followed by agroforestry management.

"In forest agriculture there is no need for more salts, there is no mineral deficiency of any kind." The forest fertilises itself by producing humus through the conversion of plant and animal waste by fungi and bacteria. There is a constant circulation of mineral material absorbed by the trees and a constant addition of new mineral material from the vast reserves underground and pumped in by the roots. The supply of nutrients is automatic and is provided by humus (organic fertiliser) and soil (mineral elements) (Howard, 2007).

Pests are less abundant in agro-forests because the species, especially the more specific ones, have great difficulty in locating and remaining on the host species when the plants are dispersed, which can occur due to chemical and visual interference in the localisation signals of the host plants (Altieri, 2002). Howard (2007) talks about the self-regulation of forests, referring to diseases: "Diseases can be found in forest plants and animals, but they are never major". Trees influence the other components of the system through their canopy, which interferes with solar radiation, air movement and precipitation, intercepting rain and modifying water particles, which reach the soil as droplets; and also through the root system, which fills large volumes of soil and exploits mineral reserves by recovering leached nutrients and making associations with nitrogen-fixing bacteria and mycorrhizae. These nutrients are then deposited on the surface. According to Pereira *et al* (2006) "for soils, the presence of trees and shrubs can induce beneficial effects such as: contribution of organic matter;

biological fixation of atmospheric nitrogen; addition of nutrients via run-off and precipitation through the trunks; reduction of soil losses; nutrient cycling; improvement of the soil's physical properties; development of soil biota and improvement of the microclimate (shading, windbreaks, etc.).

Successional agro-forestry systems are a strategy for recovering degraded areas, helping to restore soil fertility in production units. According to Peneireiro (1999), degraded areas can be recovered through agro-forestry, with a significant improvement in soil fertility and the activity of native fauna and the re-establishment of hydrological cycles, including the return of watercourses to properties. Trees are fundamental in the recovery of ecological functions within an ecosystem, and the ecological interactions between plants and animals are intensified with the planting of trees in degraded ecosystems (Campelo *et al*, 2006).

According to Altieri (2002), Agro-Forestry Systems incorporate four essential characteristics:

- Structure,
- Sustainability,
- Productivity.
- Socio-economic/cultural adaptability and are important for family farming in small areas of developing countries, since..:
- It provides areas that are better able to provide a diversified and nutritious diet;
- More efficient use of available resources and inter-specific synergism;
- Increasing the efficiency of land use and increasing the diversity of products and production throughout the year, generating income and labour at all times of the year.

In this way, Agro-Forestry Systems can help family-based farmers make their production systems more sustainable by understanding the direction to take and looking for the best Agro-Forestry System to implement in each location and the way to improve it.

An Agro-Forestry System is, according to Young (1991), a specific example of Agro-Forestry practices found in a locality or area, according to its biological composition and arrangement, technological level of management and socio-economic characteristics.

In Angola and throughout the tropical world, there are favourable environmental and socio-economic conditions for different land use systems, from intensive monocultures at one extreme to the maintenance of natural cover at others. One of the important points is the formation of more stable ecological systems, with less input of external resources and greater self-sufficiency. According to Hernandez & Benavides 1995, agroforestry should aim to maximise the use of radiant energy, minimise nutrient losses by the plants in the system, optimise water use efficiency and minimise surface run-off and soil loss through erosion. The use of agroforestry practices, with the aim of reducing erosion and maintaining and increasing soil fertility, is discussed in detail in Young (1991), based on the hypothesis that "appropriate agroforestry systems control erosion, maintain soil organic matter and physical properties and promote efficient nutrient cycling".

Other objectives include:

- Increased ecological and economic durability of the system, in view of its biological architecture, including short-cycle and long-cycle plants and animals;
- Ensuring social acceptability through a sequence of daily and seasonal activities that are easy to understand, moulded according to local tradition and designed to increase their effectiveness;
- It seeks to make full use of all inorganic resources and all available niches for useful plants and animals, while maximising the recycling of these resources;
- Reducing the risks for the individual farmer through a greater variety of useful plants and animal species and raising the quality of life and the environment. In order to analyse Agro-Forestry Systems, it is necessary to have a clear concept of the wider spatial and temporal relationships involved.

1.1.2. Limitations of Agro-Forestry Systems

Possible negative effects include allelopathic or competitive interference between trees and other crops, favouring diseases and pests, and damage to the quality of production through falling branches and fruit. These can be avoided or minimised by appropriate management and choice of species (Gomes &
Borba, 2003). The choice of suitable species can be limited on low-fertility soils, although many trees thrive on these soils, the same is not true of annual crops. There are laws restricting the felling of some native tree species.

Agricultural crops and/or pastures (animals) can compete with the tree species(s) for nutrients, space, solar energy and soil moisture and can reduce crop yields. However, this can be minimised by choosing trees with a deep root system to avoid competition with shallow-rooted crops, managing pruning, selecting components with different degrees of resource requirements, etc.

- Risks of damage during cultivation and harvesting: cultivation and harvesting operations must be planned and carefully executed, especially for high-value species systems.
- Allelopathy: seed germination and plant growth can be inhibited by the release of natural compounds (tannins, alkaloids, phenolic compounds, terpenoids, etc.) from roots and aerial parts to other plants. However, positive allelopathy can occur, potentiating the development of the components.
- Habitat or alternative hosts for pests: when close to other crops, tree species can provide a habitat for pests of all kinds. Some tree pests also affect agricultural crops and vice versa.

1.1.3. Characteristics of desirable forest species for the Agro-Forestry System

Forest species with attributes that are suitable for Agro-Forestry Systems are those that have multiple functions, i.e. forage species, nitrogen-fixing species, species with a deep root system to reduce competition with agricultural crops in the surface layers, species whose foliage is suitable for soil protection, species with a high regrowth capacity, with high production, that provide little shade.

Martinez, 1989 describes the *Acacia albida* species, which he classifies as being almost perfect for the Agro-Forestry System. During the dry season, this species keeps its leaves and loses them during the rainy season. This species is therefore suitable for intercropping with agricultural crops and pastures because it creates shade at the time when the crops need the most protection and allows full luminosity at the time when water is available in the soil for the agricultural crops to grow. Another characteristic of this species is its very deep root system, which reduces competition for nutrients. Its leaves and fruit are also top quality fodder, which makes it suitable for use in the silvopastoral system. Each species has a certain area that it needs to carry out its vital activities. The characteristics that a tree must possess in order to be desirable in an Agro-Forestry System are closely related to the aspects that favour the growth and development of the crops that are associated with the system. There doesn't seem to be a ready-made model; for each situation it is necessary to test the various types of intercropping in order to conclude which gives the best result.

1.3. Classification of Agroforestry Systems

According to their spatial structure:

- Drawing in time,
- Relative importance,
- Function of the different components,
- Production objectives and predominant socio-economic characteristics,

As for its composition:

- Agro-Silvicultural Systems (trees+crops);
- Silvopastoral systems (trees + animals);

- Agro-forestry systems (trees+crops+animals).

The current classification of SAFs is that adopted by ICRAF and the Centro Agronomico Tropical de Investigation y Ensenanza (CATIE) (OTS/CATIE, 1986) and by the Rede Brasileira Agro-Florestal (REBRAF), which is based on the type of components included and the association between them. **This classification is descriptive**: the name of each system indicates the main components, gives an idea of its physiognomy and main functions and objectives, and is therefore more didactic. Systems are classified at a first level simply as sequential, simultaneous or complementary.

- **.3.1 Sequential agro-forestry systems**

a) Improved/silvo-rotating system

Itinerant agriculture has always been practised as a traditional form of land use, and consists of felling, using the wood and burning the forest in small areas where annual subsistence crops are grown for a period of 2-3 years. The area is then abandoned, but may first be grazed, and then left fallow to regenerate to the stage of scrubland, where the cycle begins again.

The species can be introduced at the beginning or during the uniform or dense enrichment phase (Dubois, 1989). Logically, this system should be planned within a regional land use context and never as a justification for misusing land with scrubland, avoiding legal problems and providing an alternative source of income for farmers.

The organisation of community nurseries can also be part of the system. Similarly, Brienza (1986) describes a silvo-rotational system, which would be the third case, associated with cycles of itinerant agriculture, and emphasises the planting of timber forest species before the plot is abandoned. After a few years, the regenerated vegetation is cut down again for new annual crop cycles, but the commercial trees from natural regeneration and those planted are kept. With the succession of new cycles, the trees are managed and cut down at the end of the rotation.

b) Taungya" system

The term "Taungya" has its origins in Burma, meaning "hillside crops", and was originally used to designate the planting of trees in areas of shifting agriculture. Today, the term is used to refer to any combination of crops during the early stages of establishing tree plantations, where the main objective is the production of wood. This system is recommended for small farmers who have land for timber production but need to reduce establishment and maintenance costs, and for land that is forest-orientated but not degraded or steeply sloping (Beer *et al.*, 1994).

The interactions that exist in this system include interference between the components (competition, allelopathic effect) and the shading of the trees by the crops (Oliveira, 2004). Competition depends on the species chosen, density and type of management. Excessive shading by the trees, after a few years, determines the end of the Agro-Forestry System and the beginning of the pure forest plantation, the duration of which will depend on the species and density of the plantation. With the future management of the forest stand, new associations can be made with crops or fodder after thinning.

As for suitable species, some characteristics are desirable, according to Beer *et al.* (1994):

- Rapid apical growth;
- Quick closure of the canopy;
- Tolerance to shade and competition during the first year;
- Have a good stem shape;
- Natural pruning;
- Shallow shade;
- Small cupboard;
- No allelopathy;

- Deep root system.
- **.3.2 Simultaneous Agroforestry Systems**

In this group (figure 1), the agricultural and forestry components have a direct interaction, since they are on the ground at the same time throughout its duration.

a) Mixed home gardens: Like migratory agriculture, home gardens are a very old agro-forestry practice, used to provide for the basic needs of small families or communities, with the occasional sale of surplus produce. They are characterised by their complexity, with many layers and a wide variety of trees, crops and domestic animals, producing fruit, vegetables, fibres, wood, medicinal and aromatic plants, chickens, pigs and others throughout the year.

According to OTS/CATIE (1986) the main characteristics of this system are,

- The need for few inputs and constant production capacity;
- The need for labour is staggered throughout the year and concentrated in the family;
- Little economic demand and great resistance to market fluctuation and insecurity;
- They are the most similar to natural ecosystems and have high productivity per unit of land area.

The main limitation to its expansion is the scarcity of land for each family of small farmers to have their own garden.

The management of these systems includes the use of domestic organic waste and organic composts; the use of green manures of annual plants, mulch and nitrogen-fixing and firewood-producing plants; the manual control of weeds that are left as mulch; pest control is minimised by diversity and the use of resistant varieties. Another important aspect of the management of these systems is the time of year when production peaks, which influences the existence of production surpluses. One management practice also consists of selectively cutting down species for firewood or fence posts, opening up clearings that can be cultivated with annual plants.

b) Trees in association with perennial crops: In this system, shade trees are combined with shade-tolerant perennial crops, as is the case in the traditional coffee production system in Costa Rica and other tropical countries, associated with woody species, legumes and fruit trees such as mangoes, citrus fruits, coconuts and bananas. Other perennial species traditionally planted in this type of consortium are cocoa, bananas, coconuts and black tea.

The best Agro-Forestry System model for many countries in Asia, Africa and Latin America is probably the "*Taungya*" type. In countries such as Thailand, Kenya, Nigeria, Cambodia, India, the Philippines, Malaysia, Puerto Rico and some parts of Brazil, the system is already common practice among farmers (Buza, 2006).

Management practices, in addition to the appropriate choice of trees for shade (strong root system; small leaves; wind-resistant canopy; easy reproduction by cuttings; good regrowth capacity, among other characteristics), include thinning, pruning and trimming of the aerial part, mainly to influence the harvest of the perennial crop (quantity and quality). An example of this type of system is the "cabruca", a traditional cocoa growing system in the south of Bahia, in which seedlings are planted in the shade of native forest trees, after the latter have undergone thinning of their understory.

c) Trees in association with annual crops "*Alley cropping*". This group includes systems of consociation between trees and shade-tolerant annual species in different arrangements. The most common is the *Alley cropping* system, which associates rows of trees with strips of annual crops that may or may not be shade-tolerant. In this group, the interactions between the components are similar to the previous case. Row crops were developed in Nigeria and are very potential practices for all tropical regions, especially in areas with fertility problems or sloping land. Single or multi-stratified rows of trees are planted between wide strips (6-8 m) where maize, beans, cassava, soya, cereals and other annual plants are grown. Nitrogen-fixing leguminous trees associated with mycorrhizae are generally used, such as Acacia, Sesbania, Leucena, Gliricidia, Calliandra and Prosopis. The trees are planted first, in single or double rows about 0.5 m to 1 m apart.

In many cases, cuttings are planted directly in the rainy season, or seedlings produced in nurseries. The shade produced by the dense canopy of the seedlings quickly eliminates weeds and reduces maintenance costs (Assad & Almeida, 2003). Annual crops are planted simultaneously or in the next rainy season; in the latter case, after pruning the aerial part of the trees, followed by incorporation of the residues into the soil. Trees protect the soil and the crop in the dry season and are pruned annually after the agricultural harvest, for firewood and incorporation of the crown material into the soil and/or use as fodder (OTS/CATIE, 1986). Pruning trees is one of the most important aspects of managing these systems. Depending on the height of the agricultural crop and the use to which the forest product is to be put, trees are subjected to shallow cutting at 10-15 cm from the ground, or at 0.5 m to 1 m from the ground (Miranda & Valentim, 1998). Cutting at greater heights increases the ability to sprout and produce future biomass, but reduces the possibilities of utilising the trees for firewood. Alley crops can be designed to control erosion on sloping land, favouring natural terracing or reinforcing mechanically constructed embankments. In this case, it is important to choose species with a very branched and deep root system.

Fig.1-Type of Simultaneous Agro-Forestry System on a demonstration farm.

Source: Weber. 1986

Figure 2 caption:

I. Scattered trees	5. trees lining the paths	9. slope stabilisation
3. border	7.Shadow	11. Trees as living barriers
4. living fence	8. windbreaker curtains	12. Controloda erosion (ravines)

- **.3.3 Agro-Silvopastoral or Silvopastoral Systems**

They are associations of timber or fruit trees with animals and their feed, with or without the presence of annual or perennial crops (Ribeiro & Freitas, 2007). They can be practised at different levels, from large-scale forest plantations, where animals are introduced for grazing, to animal husbandry as a complement to subsistence farming systems. The main interactions in the system are:

a) The presence of the animal component can change and accelerate nutrient cycling in some respects;

b) In the case of high animal loads, there can be problems with soil compaction, which affects the growth of trees and forage plants;

c) The food preferences of animals can affect the composition of forests;

d) The trees provide a more favourable microclimate for the animals, increasing production;

e) Animals can participate in the dissemination of seeds, which cheapens implantation systems;

f) Trees can increase the forage value of the area by providing leaves and fruit, especially during the dry season.

Somarriba (1995) describes a common system in Central America of associating guava trees with pastures. Guava fruit is fed to cattle at a rate of 11kg/animal per day, and the trees spontaneously establish themselves in the pasture, forming almost pure forests associated with the pasture. These trees can be used for firewood and fruit, and also as matrix vegetation for future enrichment and recovery of degraded pastures. Manure slabs containing guava plantlets can be distributed along fences and then thinned to a height of 10-20 cm, and then managed to provide an excellent living fence (Carvalho, 2006).

Oviedo *et al* (1994) also describe a system for producing milk from goats for domestic consumption in a total confinement system, fed on trees and shrubs with a high nutritional content. This system aims to provide an alternative to cow's milk production on very small properties or on very hilly terrain, and allows for the production of 1 to 2 litres of milk per day per animal with only 1.5 hours of work between all activities. This system requires an area of 700 to 1,400 square metres per animal, 70% of which is trees and 30% mowed grass.

Another case is when the main objective is to produce forestry, latex (rubber) or fruit, and grazing is intended to help control invasive plants and reduce the cost of forestry (Favareto & Brancher, 2005). The association can begin when the trees are old enough not to be damaged by livestock, or after thinning commercial plantations of trees with longer cycles, such as Pinus and native timber. In this case, possible animal damage to the plants should be considered, as well as excessive shading or allelopathic effects that greatly reduce forage production.

1.4. Application of Agro-Forestry Systems. Different situations

- **Areas that have been deforested and degraded** by the felling and burning of trees, which favours the emission of gases such as CO_2, the exposure of the soil directly to rain, which causes erosion and silting up of rivers, imbalances in flora and fauna, with consequent biological impoverishment. These areas can be improved and/or recovered by applying agro-forestry practices such as the taungya system, sequential crops, improved fallow, multi-stratum trees, multi-use species, among others (Peck & Bisho, 1992).
- **Areas eroded by rainwater**, causing soil loss, reducing its capacity to store nutrients and water, causing high levels of soil run-off and soil compaction (Santos, 2005). These degraded areas can be recovered through the use of agro-forestry practices such as living barriers, slow land formation for agricultural use, gully stabilisation, alley crops, trees on contour lines, among others.
- **Areas of low fertility and poor drainage** generally cause losses of organic matter and nutrients, especially nitrates, and physical impediments to root development, with reduced tree growth and nutrient deficiencies in annual crops (Khatounian, 2001). They can be recovered with agro-forestry practices such as row crops, strip crops, forest foliage as a source of fertiliser, trees around agricultural crops and pastures, among others.
- **Dry (arid) areas with hard soils** that have difficulty storing water and nutrients; high temperatures that affect evapotranspiration and the water table. They can be recovered using agroforestry practices such as living barriers, windbreaks, living fences, trees around crops and pastures, among others.
- **Hillside areas (steep slopes)** are generally areas devoid of forests, with a high erosion rate and difficulty in forming permanent soil cover. They can be recovered by using agroforestry practices such as rows of trees on land, strip cropping and living barriers (Otarola, 1999).
- **Fallow areas and/or marginal areas** of low ecological and economic value. They can

be recovered by agroforestry practices such as improved fallow and multi-stratum trees.

- **Degraded pasture areas with poor vegetation cover**, exposing the soil to the damaging effects of water and wind erosion. They can be recovered through agro-forestry practices such as pasture afforestation and forage banks, among others.

1.4.1. Aspects to take into account when establishing Agro-Forestry Systems.

According to Assis (2009) it is necessary to follow the steps below before establishing Agro-Forestry Systems.

- Determine the geographical limits of the area and the objectives of the characterisation.
- To collect physical, biological and socio-economic data on the region.
- Survey the characteristics of existing production systems.
- Determine the problems, needs and opportunities in the area.
- Analyse the above data in order to determine whether the use of Agro-Forestry Systems is feasible or appropriate for that region.

In the case of this study, these steps were followed through the surveys carried out.

1.5 Concepts related to good practice

It can be said that of all the natural resources on the planet, soil is one of the most unstable when it is modified, i.e. when its protective layer is removed. Soil is a finite resource.

1.5.1. Composting

Production of organic compost (fertiliser) made from humidified organic material obtained from the transformation (biological decomposition) of organic waste (leftover crops, fruit, vegetables, animal waste, etc.) by the microbial action of the soil (Machado *et al*, 2006).

- It is used to enrich poor soils, improving their structure and providing good fertility;
- Increases plants' ability to absorb nutrients (macro and micro) by providing substances that stimulate their growth;
- Facilitates soil aeration, water retention and reduces water erosion;
- It acts as an inoculant for the soil, accumulating macro and microorganisms (fungi, actinomycetes, bacteria, earthworms and protozoa) that are natural soil builders.

1.1.1.1. Material to use for composting

- Animal waste (manure from chickens, cattle, pigs, etc.);
- Unmarketed peelings, fruit berries and stones;
- Crop residues (rice husks, maize straw, dried bean pods, dried coffee husks);
- Leaves and branches of cassava, banana;
- Sawdust;
- Grass clippings;
- Ashes;
- Black earth.

1.5.2. Crop rotation

It consists of alternating plant species every year in the same agricultural area. The species chosen must have both commercial and soil recovery purposes.

Plot 1 - Leafy vegetables such as lettuce, spinach and cabbage, but also potatoes and tomatoes. These need a soil rich in nutrients, particularly nitrogen.

Parcel 2 - This includes onions, garlic and other root vegetables such as radishes, carrots, beetroot and turnips, which are able to draw nitrogen from deeper down.

Plot 3 - This is where the grain vegetables are. Beans, peas and also fruit vegetables such as courgettes.

Plot 4 - This plot is left fallow. Here you will only sow "green manure", i.e. plants such as clover, lucerne and spontaneous vegetation, which will enrich your land.

SIMPLIFIED REPRESENTATION OF A "CROP ROTATION" SCHEMEWi.^

1.5.2.1. How to route

In the second year, place the crops from plot 1 on plot 2, the crops from plot 2 on plot 3 and the crops from plot 3 on plot 4. Then rest plot 1. The idea is never to repeat the same crops on the same plot. It's also important to enrich the soil with manure or compost. Crop rotation has the following advantages:

- **They increase soil fertility** - if the crops and the rotation period are more suitable;
- **They reduce soil impoverishment** - alternating crops means that the different layers are exploited in depth by roots with different characteristics;
- **They make it easier to control pests, diseases and weeds** - by alternating crops with different characteristics.

1.5.3. Crops in travelling curves

It is the installation of crops according to the lines of the land that have the same altitude (elevation).

With this arrangement, soil erosion is avoided because rainwater is directed with little speed towards the respective water lines.

(In order to keep the plant on high ground so as to avoid erosion and flooding of the plant, it is necessary to make the plant in curves following the relief of the land. This technique is called contour lines). In the countryside, peasants traditionally use vipangas (local name) for this purpose.

Expeditious methods for determining contour lines:
plumb line and triangular square

Fonte: Götsch, 1995

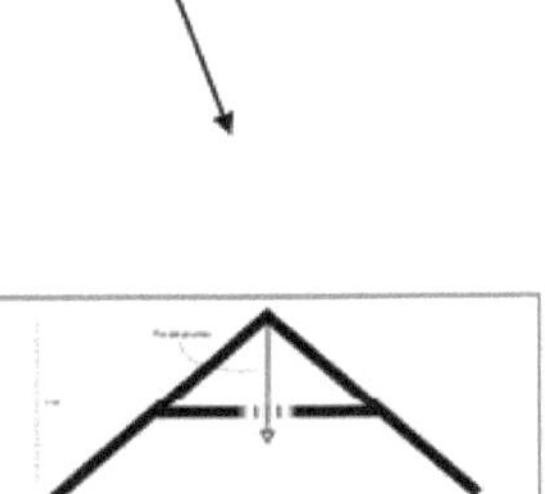

Fonte: Loedeman, 2005

Source: Loedeman, 2005

CHAPTER 2

2. MATERIALS AND METHODS

2.1. Characterisation of the study area

The province of Huambo is located in Agricultural Zone 24, according to the classification of the Angolan Agricultural Survey Mission (MIAA), the Central Plateau, and is largely situated at an altitude of over 1,500 metres. The climate is of the humid tropical type modified by altitude, characterised by two well-defined seasons: the rainy season, which lasts from late September to April, and the dry season, which lasts from early May to September. During the rainy season there is often a short dry period (small cacimbo), which in recent years has lasted more than two weeks. Rainfall values vary between 1,100mm to the S-SW and 1,400mm on the central-western plateau (Diniz, 1973). The predominant type of vegetation is open forest, commonly known as panda forest.

The province's hydrographic network is quite significant, with the Queve, Cunene and Cubango basins standing out. These basins offer enormous energy and irrigation potential, as well as rich fishing resources.

The province of Huambo is one of Angola's regions with serious environmental problems, mainly related to soil degradation and the destruction of forest areas. The problem with Huambo's soils is essentially structural; they are mostly ferraHtic, with low mineral reserves and low organic matter content, which explains their low fertility, contrary to the common-sense opinions that Huambo's land is fertile. In this respect, it's worth emphasising that scientific studies carried out during the colonial period had already concluded that the Central Plateau region was not, and did not have the conditions to be, Angola's celebrated granary.

In addition to this characteristic aspect of Huambo's soils, there is also the intensive use of cultivated areas in the face of increasing demographic pressure, as noted in the previous point.

ESBOÇO DE LOCALIZAÇÃO

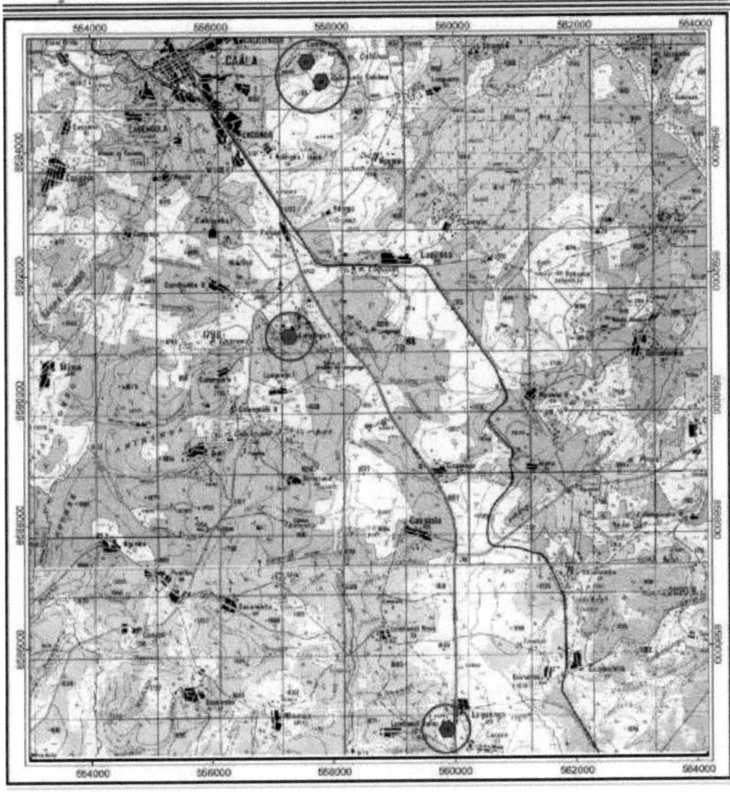

DATA 12/06/2012	*ALDEIAS DE CALUE, CAMBONGUE, LUNGONGO E CACACA*	
ESCALA 1/80.000	LOCAL: *Comuna Sede* *Município da Caála* *Província do Huambo*	
FOLHA Nº 256		

Fig. 2 - Localisation map drawn up by IGCA-Huambo

The municipality of Caala is part of the province of Huambo in the west. It covers an area of 3,680 square kilometres and its population is estimated at approximately 191,972 inhabitants. The villages in the municipality of Caala where the surveys were carried out are the villages of Calue and Cambongue, located in the south-east and about 13 kilometres from the municipal seat, with an estimated population of 658 families, the village of Lungongo concentrates about 157 families, is located to the south of the seat municipality, 15 kilometres from the seat municipality (Caala), bordered by the Lungongo and Kususu-Kosaba rivers. The village of Cacaca is located to the south of the municipality, with around 130 families and is 25 kilometres from the municipality (Fig.2).

The work is centred on the villages of Calue and Cambongue due to the degree of degradation they present, which merits urgent action, especially in the village of **CALUE.**

2.1.1. **Calue Village**

The rate of deforestation has reached extreme levels. Soils devoid of vegetation are completely at the mercy of erosive agents. The presence of ravines is evident, and if they are not stopped, they will continue to advance, making it increasingly difficult to recover from them (figure 3).

Fig. 3 - Aspect of the village of Calue. Source: MIGUEL, Agnelo (2012)

Calue is a village whose Umbundu name means "Stone" in Portuguese. This stone is located in the village of the same name. It was founded in 1947. The village is run by the traditional power, made up of a soba, Mr Laurindo, his deputies Mr Abel Nguluka and the "akuendje vokovambo", the village poHcias. Five sobas passed through Calue before the current one, Mr Kapaco, Mr Armando, Mr Kandjeke, Mr Inancio and Mr Lundo respectively. There is only one Catholic church and one primary school, which is close to the neighbouring village (Vikassa).

The community depends on agriculture, mainly for subsistence. The main crops they produce there are: maize, beans, sweet potatoes and reindeer potatoes. Horticultural crops are produced very sparingly due to a lack of water.

The villages of Calue and Cambongue are located in one of the areas of the municipality of Caala characterised by profound forest degradation, with almost no natural forest. Given this level of degradation, firewood and other forest resources are scarce and to obtain them the inhabitants of these villages are forced to travel more than fifteen kilometres to other locations. Meanwhile, the process of forest degradation has intensified in recent years with the increase in cattle and the area does not have enough areas for grazing, which is causing overgrazing that could, in the not too distant future, have harmful implications for the functioning of local ecosystems, some of which are already evident.

Deforestation combined with the effects of soil degradation require the adoption of production system models that favour more sustainable management of forest resources. The agroforestry model is the production system that best suits the agroecological context of these communities.

2.1.1.1. Access to Land

- Heran^as
- Letting
- Loans
- Purchase

For the time being, there are no land conflicts. That's why the community is concerned with delimiting and legalising its land.

The village of Cambongue presents a similar situation as they are neighbouring villages.

2.2. Materials used

I. To carry out the enquiries:

- Enquiry
- Computer
- Camera
- Ballpoint pen

II. For soil collection and analysis

- Drilling rig
- 500g polythene bags
- Reagents
- Greenhouse
- Pipettes
- Potentiometer

2.3. Model enquiry

The survey template in Annex 1 contains a short introduction to the objectives and then questions on the following aspects:

1. Type of mining, practices used and adherence to Agro-Forestry Systems (AFS);
2. Water and related issues;
3. Forestry and related issues.

A total of 131 surveys were carried out. The approach was as follows: greet the highest traditional authority, members of local government and farmers. Then, after a brief conversation with the aforementioned authorities, the objectives were explained in the national language of Umbundu.

2.4. Different types of mining

According to the physiographic conduits (landforms and their implications for the evolution of forms in different geosystems) prevailing in the area, we can make a distinction:

Mongongu or Ongongo - which corresponds to ploughing from above on forest land or woodland (figure 4);

Fig. 4 - Localisation of an Ongongo or high-grade mine

Onaka or Enaka - at the bottom of the valley, which corresponds to lowland ploughing (draining and conserving soil moisture by controlling the phreatic lengol);

Ombanda - on the edge of the lowland, where drainage and water management are easier (figure 5);

Fig. 5 - Location of an Onaka and an Ombanda

Otchumbo - corresponds to the backyard peak enriched by the organic accumulations around the house (figure 6).

Fig. 6 - Location of an Otchumbo, surrounded by avocado trees. Inside we can see a reindeer potato plantation.

Each of the respondents has different types of mines from those mentioned above. This aspect was one of the questions considered in the survey, as well as the area of the respective farm.

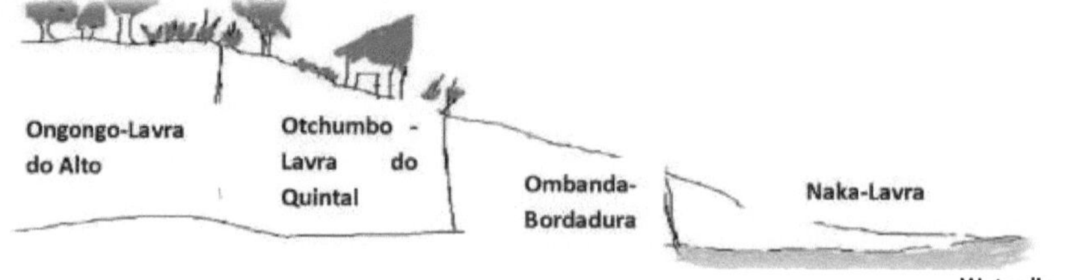

Fig. 7 - Catena utilisation scheme in Huambo (Fernando Pacheco, 2005)

2.5. Methods

2.5.1. How the data is processed

1. Introducing the answers per person and per theme (the same criteria were used for each of the villages)
2. Based on this data, the number of answers to the question was calculated.
3. Based on these figures, the percentages were calculated taking into account the number of respondents per village.
4. With these values, we determined the means and their standard errors.

The graphs and tables were drawn up on the basis of the data obtained as described in points 1 to 5. With this data we were able to interpret the socio-environmental phenomena (farming systems, maintenance of soil fertility and the state of natural resources) in the study areas (Calue, Cambongue, Lungongo and Cacaca villages).

2.5.2. Collection and preparation of soil samples

The soil was sampled in a zigzag pattern with a probe at a depth of 020 cm. The pH was then determined at laboratory level. To do this, the upper part of the soil was cleaned of weeds, dried leaves, etc.

Once the various sub-samples had been collected, they were bagged and identified and transported to the test site, where they were dried in an oven. After the drying period, the samples were sieved through a 2 mm mesh and homogenised.

2.5.3. Determining soil pH

The pH of the soils was determined in water and potassium chloride as described below:

1. Weigh 10g of soil in triplicate for each of the three 50ml precipitation cups;
2. Add 25 ml of distilled water and 25 ml of normal Kcl solution (1 N);
3. Calibrate the potentiometer using the cap solutions pH 4.0 and pH 7.0 or pH 9.16 (or pH 9.0) and operating according to the instructions that come with the device; i.e. before calibration, switch the device on and adjust it to the level of the cap solutions;
4. Stir before immersing the previously washed and dried electrode (with filter paper) in the soil suspension, continuing to stir after the value has stabilised;
5. Recording pH values;
6. Once the recordings have been made, the device is switched off, the electrode is rinsed with distilled water and dried. The electrode is then placed inside a lid containing a solution of Kcl to better preserve it.

CHAPTER 3

3 RESULTS AND DISCUSSION

The results of the various topics covered by the surveys are presented and discussed on the following pages. Some of them will be included in the annexes because, although they are important to the subject in general, they were considered less pertinent.

2.6. Type of mining

2.6.1. Distribution of women and men surveyed by village

The percentage of women is higher than the percentage of men, with the exception of the village of Cacaca, where the percentage of women is 46 and the percentage of men is 54 (Table 1).

Table 1 - Distribution of women and men surveyed by village

Village	N° Women	% M	N° Men	% H	Total
Calue	36	60	24	40	60
Cambongue	8	53	7	47	15
Lungongo	16	53	14	47	30
Cacaca	12	46	14	54	26
					131

This data leads us to reflect on the importance of women in the introduction and adoption of new production systems, as has happened in Kenya, Malawi and Zambia with Agro-Forestry Systems (Christina *et al*, 2002). According to these authors:

1° Farmers plant several small plots of trees,

2° They wait three to four years to see the results of each plot.

Because this "improved fallow" cycle takes so long, adopting or adapting this technology takes much longer than introducing an improved seed or a new fertiliser.

This is especially true for women farmers, especially those who are responsible for the household, whose lack of adult family members to help with the work in the fields causes them serious money and credit constraints.

This scenario is very similar to what is happening in Huambo today. In the villages there are many women and children, to the detriment of the number of men who are moving to the urban centres in search of other ways of earning a living. Agriculture has low yields and does not guarantee the subsistence of the household.

Table 2 - Relative % of the different types of ploughing by the inhabitants of Calue village

Type of ploughing in Calue		Relative %
Otchumbo	4	6,67
Onaka	1	1,67
Ombanda	0	0,00
Ongongo	0	0,00
Otchumbo+ Onaka+ Ombanda +Ongongo	36	60,00
Otchumbo+ Onaka+ Ongongo	1	1,67
Otchumbo+ Ongongo	9	15,00
Otchumbo+ Ombanda +Ongongo	1	1,67
Otchumbo+ Onaka+ Ombanda	4	6,67
Otchumbo+ Onaka	4	6,67

Analysing Table 2, it can be concluded that 60% of the respondents have all four types of plough: Otchumbo, Onaka, Ombanda and Ongongo. This result shows that farmers distribute the degree of harvest risk in this way. Those with only Otchumbo and Ongongo account for 15 per cent.

And those with Otchumbo have the same percentage as the group with Otchumbo+Onaka+Ombanda and Otchumbo+Onaka, i.e. around 7 per cent. It is noteworthy that Otchumbo is present in the most representative ensembles. This reveals the importance of this type of ploughing in the context of the soil management and deforestation observed in the village of Calue. In this region there are no longer any areas with forest that would allow them to open new plots.

In fact, the content of organic matter in otchumbo is likely to be much higher than in other ploughlands because, due to the proximity of residences, domestic waste (husks, firewood, maize straw, dried bean pods, ashes), the existence of domestic animals and the waste they accumulate (animal droppings) all contribute to the fertility of these soils.

According to Gotsh (2007), 10 million hectares of productive agricultural land in the world are affected by desertification every year.

These percentages are from areas that could produce much more, but due to the process of soil degradation, it is progressively decreasing. Land degradation (degradation of soils and water resources, degradation of vegetation and biodiversity and reduction in the quality of life of affected populations) in arid, semi-arid and dry sub-humid regions is the result of several factors. These include climate change and unsustainable human activities (UN, 2006). Reference should be made to the degradation of onakas in the village of Calue, since the silting of the rivers has marked expressions that involve the Onakas, contributing to the extinction of this very important type of mining.

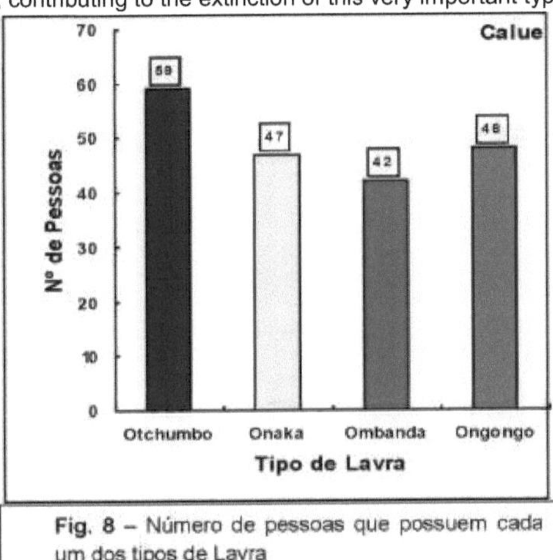

Fig. 8 – Número de pessoas que possuem cada um dos tipos de Lavra

3.1.3. Type of ploughing in the village of Cambongue

Table 3 - Relative % of the different types of ploughing by the inhabitants of Cambongue village

Type of ploughing in Cambongue		**Relative %**
Otchumbo	0	0
Onaka	0	0
Ombanda	0	0
Ongongo	0	0
Otchumbo+ Onaka+ Ombanda +Ongongo	14	93
Otchumbo+ Onaka+ Ongongo	0	0
Otchumbo+ Ongongo	0	0
Otchumbo+ Ombanda +Ongongo	1	7
Otchumbo+ Onaka+ Ombanda	0	0
Otchumbo+Onaka	0	0

Table 3 shows that 93% of farmers have all four types of plough: Otchumbo+ Onaka+ Ombanda +Ongongo. The remaining 7% is represented by the Otchumbo+ Ombanda +Ongongo combination, once again showing the presence of Otchumbo, in line with the findings for the village of Calue. The same data can be found in figure 10, where 15 people have Otchumbo.

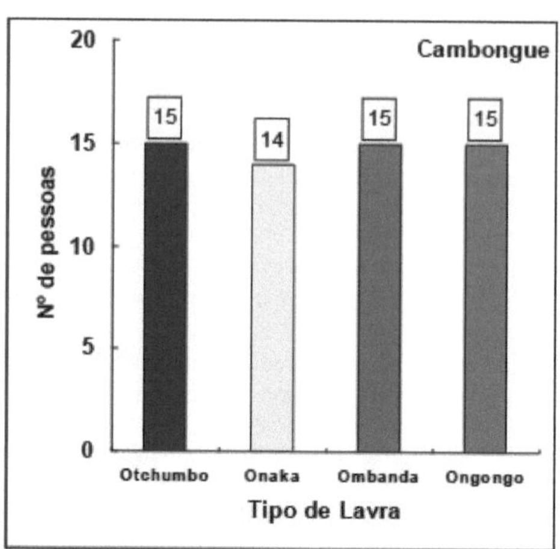

Fig. 9 - Number of people who own each type of farm

Fig. 10 - Farmers from the village of Lungongo gathered in Jango during the clarification session before filling in the questionnaires.

3.1.4. Type of ploughing in the village of Lungongo (Table 4)

Table 4- Relative % of the different types of ploughing by the inhabitants of the village of Lungongo

Type of Lavras no Lungongo		% RELATIVE
Otchumbo	0	
Onaka	0	
Ombanda	0-	
Ongongo	0	
Otchumbo+ Onaka+ Ombanda +Ongongo	30	100
Otchumbo+ Onaka+ Ongongo	0	
Otchumbo+ Ongongo	0	
Otchumbo+ Ombanda +Ongongo	0	
Otchumbo+ Onaka+ Ombanda	0	
Otchumbo+ Onaka	0	

In Table 4, 100 per cent of the farmers have all four types of plough. In this village, as in previous ones, the percentage of women farmers exceeds that of men. This is also true in terms of dynamism and openness to new methodologies.

3.1.5. Type of ploughing in the village of Cacaca

Table 5 - Relative % of the different types of ploughing by the inhabitants of the village of

Cacaca

Type of ploughing in Cacaca		Relative %
Otchumbo	0	
Onaka	0	
Ombanda	0	
Ongongo	0	
Otchumbo+ Onaka+ Ombanda +Ongongo	26	**100**
Otchumbo+ Onaka+ Ongongo	0	
Otchumbo+ Ongongo	0	
Otchumbo+ Ombanda +Ongongo	0	
Otchumbo+ Onaka+ Ombanda	0	

3.2 Practices used

1. **Soil condition**

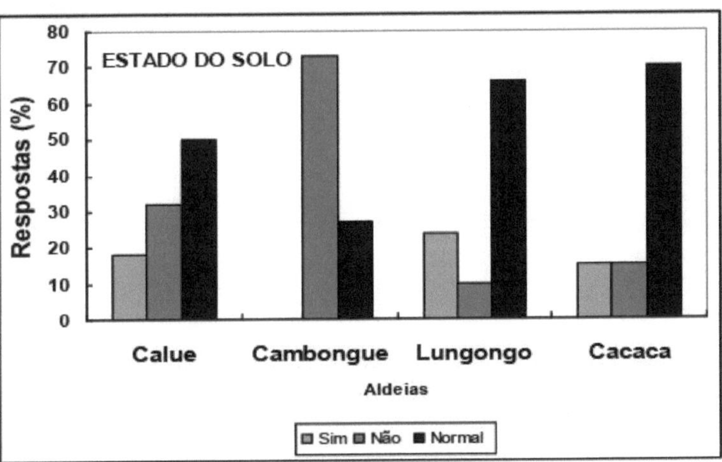

Fig. 12 - Responses from farmers in the 4 villages regarding the state of the soil (values converted into %).

The following can be concluded from analysing figure 12:

With the exception of the village of Lungongo, the number of farmers who answered that the soil is not good is higher than those who answered yes.

In this village, 72 per cent of respondents do not fallow land and 66 per cent open new plots in the forest.

In Cambongue, which is in the project's implementation area, 73 per cent of the people surveyed answered that the soil is not good. No-one answered Yes.

It should be remembered that in the village of Cambongue the plantation areas are larger than in Calue and that around 80 per cent of the farmers have all four types of plough.

Also in the village of Cambongue, 20 per cent of respondents answered that they cleared a new plot of land in the forest and 53 per cent didn't set aside any land.

II. Agricultural practices

Table 6 - Good practices carried out in the 4 villages

PRACTICES	CALUE		CAMBONG		LUNGONGO		CACACA	
	YES	NO	YES	NO	YES	NO	YES	NO

Composting	39	21	7	8	25	5	18	8
Crop Route	16	43	10	5	22	8	11	15
Crops on contour lines	16	42	3	12	5	25	6	20

Composting - The village of Lungongo has the highest percentage of composting - 86%[1], followed by the village of Cacaca with 69%[1], the village of Calue with 65% and lastly the village of Cambongue with only 47% (figure 13 and table 6).

Crop rotation - The village of Lungongo has a percentage of 73, Cambongue 67, Cacaca 42 and lastly Calue with 27 (figure 13 and table 6).

Crops on contour lines - All have percentage values below 30, i.e. very low. Calue with 27%, Cacaca with 23%, Cambongue with 20% and finally Lungongo with only 17% (figure 13 and table 6).

In some villages, ploughs with areas close to 8 ha have been identified. These areas are obtained by clearing new areas of forest. Taking good practice into account, the farmers with the largest areas only compost and have 8 and 7 ha of planted land respectively.

The largest areas belong to farmers who have all 4 types of ploughing; however, there is at least one who has both lead and ongongo and occupies an area of 41250 m^2 and another with only lead with 42500 m^2.

In both cases, these are farmers who compost but don't plant on contour lines or rotate crops. In terms of percentage distribution by type of mining, the order is ongongo with 47 per cent, lead with 22 per cent, onaka with 16 per cent and ombanda with 15 per cent.

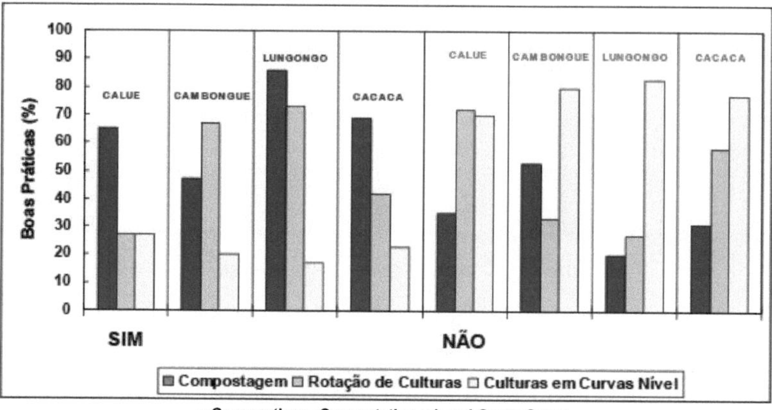

□ **Composting** □ **Crop rotation** □ **Level Curve Crops**

Fig. 13 - Farming practices of farmers in the 4 villages (values converted into %).

III. Maintaining fertility

[1] These are two villages with large numbers of cattle, pigs and goats.

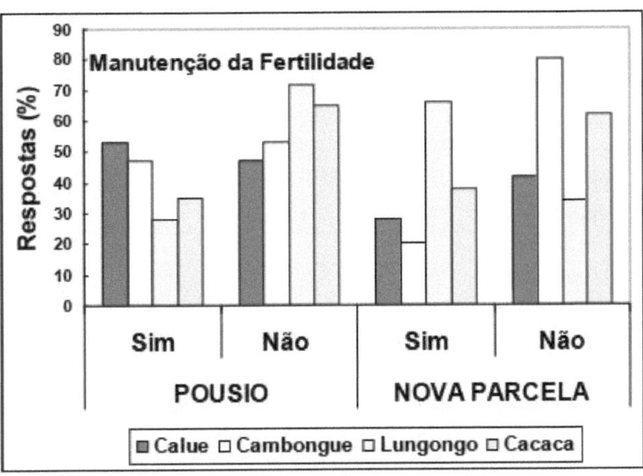

□ Calue □ Cambongue □ Lungongo □ Cacaca

Fig. 14- Responses from farmers in the 4 villages regarding the state of the soil (values converted into %).

- With the exception of the village of Lungongo, the number of farmers who answered that the soil was not good was higher than those who answered yes (figure 13).

- In this village (Lungongo), 72 per cent of respondents don't fallow and 66 per cent open new plots in the forest (figure 14).
- In Cambongue, 73 per cent of the people surveyed answered that the soil is not good. No-one answered Yes (figure 13).
- Let's remember that in the village of Cambongue the plantation areas are larger than in Calue and that around 90 per cent of the farmers have all four types of plough Figure 14 and Table 3).
- Also in the village of Cambongue, 20 per cent of respondents said that they opened a new plot in the forest and 53 per cent did not set aside any land (figure 14).

3.3. Adherence to Agroforestry Systems

Table 7 - Farmers' main fears when adopting SAFs

Main fears in adopting FAS (%)		
	Technical	Finance
Calue	82	0
Cambongue	100	100
Lungongo	18	30
Cacaca	69	100

Analysing table 7, it can be seen that, with the exception of the village of Calue, all the other villages have both technical and financial fears. The village of Cambongue has 100% technical fear and 100% financial fear. In the village of Cacaca, around 70% fear the technical aspects and 100% the financial aspects.

Chart 8 - On which land do you prefer to install SAFs?

	Your Land	Community Land
Calue	50	9
Cambongue	15	0
Lungongo	28	1
Cacaca	26	0

According to table 8, the vast majority of respondents prefer to experiment on their own land. All the farmers in Cambongue and Cacaca want to test on their own land.

Table 9 - Farmers' preferences (%) when introducing SAFs.

Onaka+ombanda+ongono+otchumbo	21
otchumbo	20
ongon	9
ombanda	9
ongon + lead	9
Onaka+ otchumbo	9
ombanda+otchumbo	7
ombanda+ongono+otchumbo	5
Onaka	4
Onaka+ombanda+otchumbo	4
ombanda+ongono	2
Onaka+ ongono+otchumbo	2

Table 9 expresses farmers' wishes regarding the type of plough where they prefer to introduce SAFs. It's interesting to see that the percentage of people who want to adopt SAFs on the 4 types of plough (21%) is the same as those who prefer to adopt them on their Otchumbo (20%). On the one hand, the distribution of risk and the observation of behaviour in different situations. On the other hand, the security of the lead, where you can monitor developments on a daily basis.

Table 10 - Farmers' receptiveness to the introduction of SAFs and preferred species (Percentage values)

			Species to be introduced					
	With good results you'll stop opening new areas (%)		Forestry (%)		Forest and fruit trees (%)		Forestry, Fruit trees, Improvers (%)	
	Yes	No	Yes	No	Yes	No	Yes	No
Calue	93	7	8	0	78	0	14	0
Cambongue	100	0	0	0	40	0	60	0
Lungongo	100	0	0	0	70	0	30	0
Cacaca	100	0	8	0	23	0	69	0

Table 10 shows that the species favoured by farmers for introduction into SAFs in the village of Calue are firstly a combination of fruit trees and forestry; forestry, fruit trees and improvers and lastly only forestry. Lungongo follows the same trend. In the village of Cambongue, they only want fruit trees + forestry. In Cacaca, the trend is fruit trees + forestry + improvement; forestry + fruit trees and finally forestry. Given the type of degradation problems that exist in these villages, the introduction of improvement crops is essential.

Table 11 - According to the results obtained with the installation of SAFs (percentage values)

	It will encourage others		You'll let it make coal	
	Yes	No	Yes	No
Calue	100	0	65	35
Cambongue	100	0	7	93
Lungongo	100	0	83	17
Cacaca	100	0	89	11

Table 11 shows the farmers' intention to encourage others to join SAFs if they are successful. When asked whether they would stop producing charcoal if the SAFs were successful. In the village of Cambongue, 7% of respondents said they would not continue, while 93% said they would continue producing charcoal. This situation stands out from that of the other villages and is an exception.

3.4. Water and related issues

According to the data collected, which can be found in the annexes (Annex 3 - Table I), we began by analysing the answers given to the various questions asked about the rain. There was unanimity in saying that the rainy season has changed compared to previous seasons and that the amount of rain has decreased. These answers are reflected in the amount of water needed to guarantee rainfed crops, which most farmers in the four villages realise is in short supply.

Annex 3 contains data on the quantity and quality of river water. We also questioned the quality and quantity of water from the cacimbas, the main source used for cooking and drinking.

Respondents say that in dry weather there are very few cisterns that have an abundance of water, as can be seen in Tables III and IV in Annex 3. As for the quality, they report that it is not of good quality, causing some illnesses, which forces them to boil the water (Table 4 in Annex 3).

Regarding the flow of the rivers, there was unanimity in saying that the quantity of water is different or lower. As for whether the water is clean or has rubbish in it, the village of Cambongue is the only one with a larger number of respondents saying that the water in the rivers is not clean and reinforcing this by saying that it has rubbish in it. Farmers in the village of Cambongue say that the water is not clean, but more respondents say that the water has rubbish in it. The same goes for the village of Cacaca.

The farmers in the village of Lungongo are unanimous in saying that the water is clean and there is no rubbish.

Table III (Annex 3) summarises the answers given as to who causes "rubbish". In the village of Calue, they attribute this rubbish to people and the wind. The same tendency applies to the village of Cambongue, which is in line with the answers given in Table 12. The farmers in the village of Lungongo also gave answers in line with the answers given in Table 12.

Most of the respondents in the four villages say that they use water from the cacimba for drinking and cooking. In dry weather, there are few cacimbas with an abundance of water, according to the farmers' responses.

In the village of Calue, most farmers say that the water from the cacimba lasts until October. The same trend can be seen in Cambongue. Lungongo and Cacaca show great difficulties with water from the cacimbas from July to October (Table V in Annex 3).

According to the results in Table V, all the farmers mention the poor quality of the water. Almost all of them boil the water to prevent illness.

3.5. Forests and related aspects

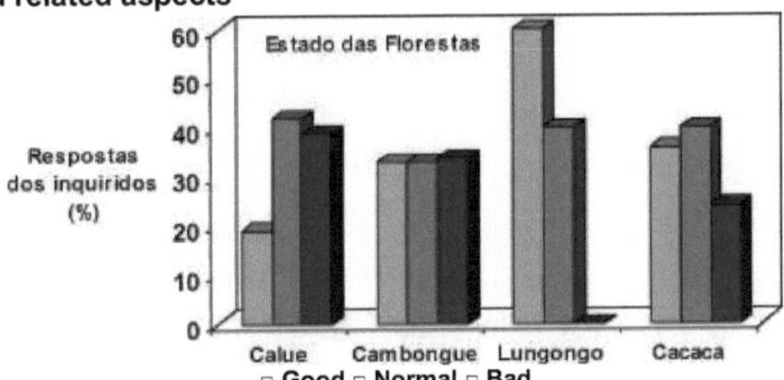

Fig. 15- Responses from farmers in the 4 villages regarding the state of the forests (values converted into %).

In the village of Calue, 19 per cent of farmers said that the state of the forests was good; Around 40 per cent answered normal and bad respectively (figure 15).
As for the village of Cambongue, the % of farmers who responded to the three situations was identical (figure 15).
In Lungongo, the majority consider the forest to be in good condition (figure 15).
In Cacaca, the responses for good, normal and poor forest condition were 36, 40 and 24 per cent respectively (figure 16).

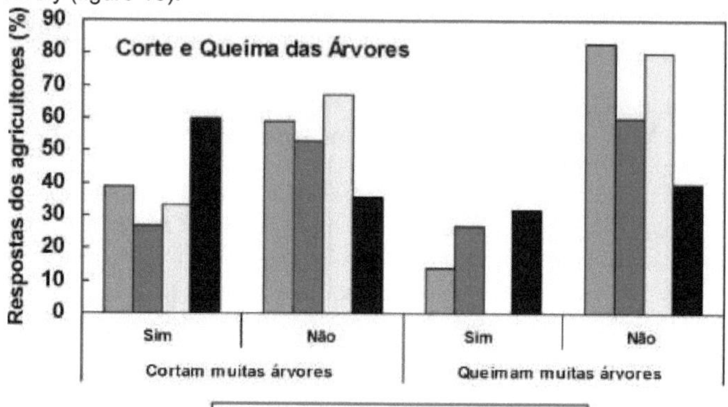

Fig. 16 - State of the Forests. Considerations made by farmers regarding the felling and burning of trees.

Analysing figure 16, which refers to the state of the forests in terms of cutting down and burning trees, in the village of Calue, 40% said that they cut down a lot of trees, while 60% of farmers said that they don't cut down many trees. In the same village, 20 per cent say that they burn too many trees, while around 80 per cent say that they don't burn them. As for the neighbouring village of Cambongue, 27% said that they cut down trees, while 53% said that they don't do it. The response to the burning of trees coincides in percentage terms.
In Lungongo, according to 70% and 80% of the farmers, they don't cut or burn very much, which coincides with the answers they gave regarding the state of the forests as shown in figure 16. In Cacaca, 60 and 32 per cent of the respondents said that they cut down a lot and burn many trees, which somewhat contradicts the answers given by these farmers regarding the state of the forests. Cutting down and burning a lot of trees certainly means that the state of the forests can be neither good (36%) nor normal (40%).

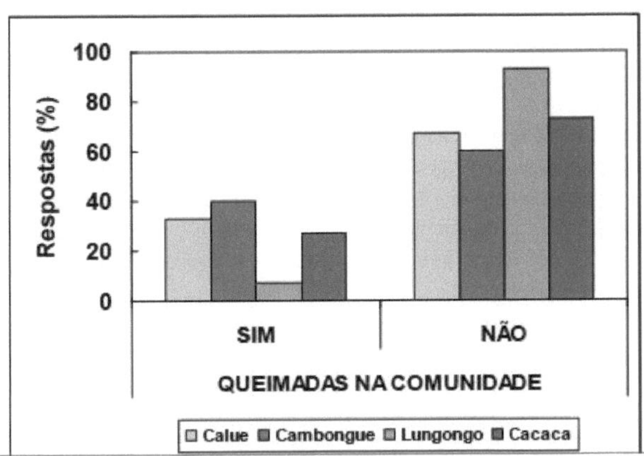

Fig. 17 - Considerations made by farmers regarding burning in the community

Figure 17 shows that 33 per cent of respondents in the village of Calue say they burn, while 67 per cent say they do not. In the village of Cambongue, the percentages are similar (40% Yes and 60% No). In the villages of Lungongo and Cacaca, 93 per cent and 67 per cent respectively say they don't burn. There seems to be agreement with the data shown in figure 16.

With regard to the trees that are closest to the village (figure 18), there is a predominance of eucalyptus, which only in the village of Cacaca is slightly surpassed by "onduko", a local name given to trees in the Miombo Forest that are suitable for making charcoal, usually *Brachystegia* spp.).

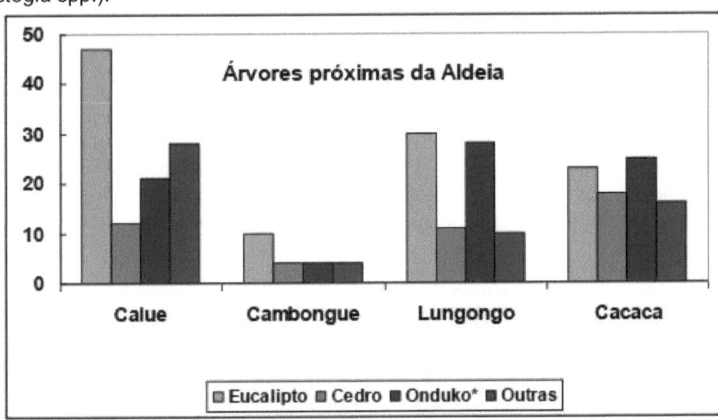

Fig. 18 - State of the Forests. Considerations made by farmers regarding trees near the village.

Table 12 - Good forestry practices (Percentage values)

	Replant the trees they cut down		Forest seedling production	
	Yes	No	Yes	No
Calue	30	70	78	22
Cambongue	60	40	100	0
Lungongo	0	100	100	0
Cacaca	0	100	100	0

With regard to good forestry practices (Table 18), all the villages produce forest seedlings: Calue 78 per cent and 100 per cent in the remaining villages. The main species produced are pine and eucalyptus.

As for replanting, the villages of Calue and Cambongue are tending to start this practice (30% and 60% respectively). This is certainly linked to the valuable work that Angola's ADRA has been doing in these villages, where it has set up community nurseries with forest and fruit species with the help of the farmers. In the villages of Lungongo and Cacaca, they only mention the production of forest seedlings, which is a source of income since it is for sale and not for replanting the trees they cut down (Table 12).

Some farmers said that, in addition to the Angolan ADRA, seedlings are also purchased through the EDA, the IDF and some unspecified non-governmental organisations.

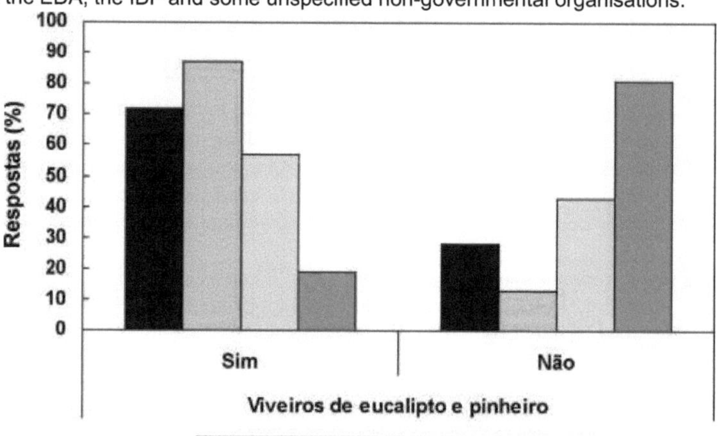

Fig. 19 - Existence of forest species nurseries (pine and eucalyptus). Percentage figures.

With regard to nurseries for natural forest species, 100 per cent of those surveyed said that they don't do this. As they know that natural regeneration of natural forest exists and have little knowledge of the concept of endangered species, they think it unnecessary to think about this work.

3.6. Climate and related aspects

Table 13 - Answers related to forests and temperature

	The forest is important for the temperature		**Fires in the community**		**The cold it was this year and the same as back in the day?**		**They used to feel colder**	
	Yes	**No**	**Yes**	**No**	**Yes**	**No**	**Yes**	**No**
Calue	100	0	33	67	2	98	30	70
Cambongue	100	0	40	60	40	60	20	80
Lungongo	100	0	6	94	6	94	10	90
Cacaca	100	0	27	73	0	100	35	65

Table 19 shows a summary in percentage values of questions relating to the importance of the forest for maintaining the temperature, fires in the community, how cold it is now compared to 10/15 years ago and how cold they feel now. The questions were formulated in this way to gauge the rural population's estimation of climate change, which is intrinsically related to high rates of deforestation, among other factors.

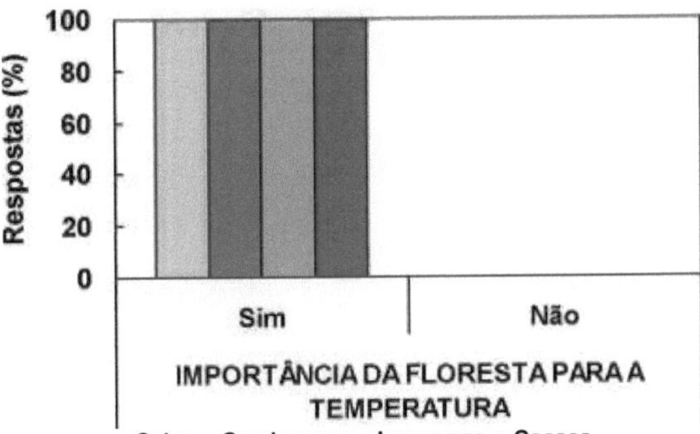

Fig. 20 - Importance attributed by farmers to the forest in relation to temperature (Percentage values).

More explicitly, we can see in figures 21, 22 and 23 that 100% of the farmers recognise the importance of the forest in mitigating the temperature (figure 21), where, with the exception of the village of Cambongue, they all say that the current cold is different (less cold) from the cold they used to feel (figure 21) and, consequently, they also feel less cold (figure 22).

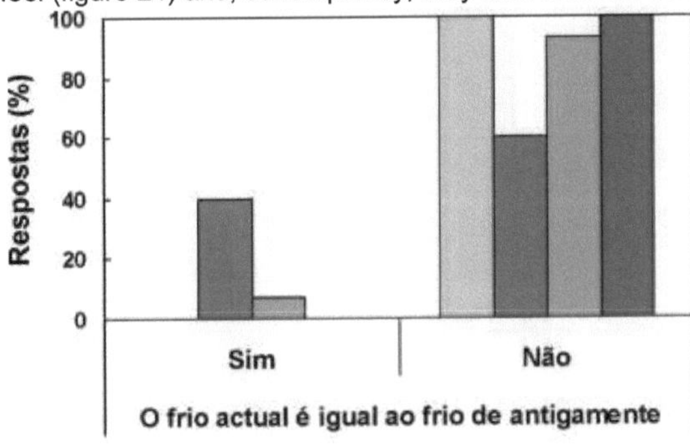

Fig. 21 - Sensitivity to climate change (percentage values).

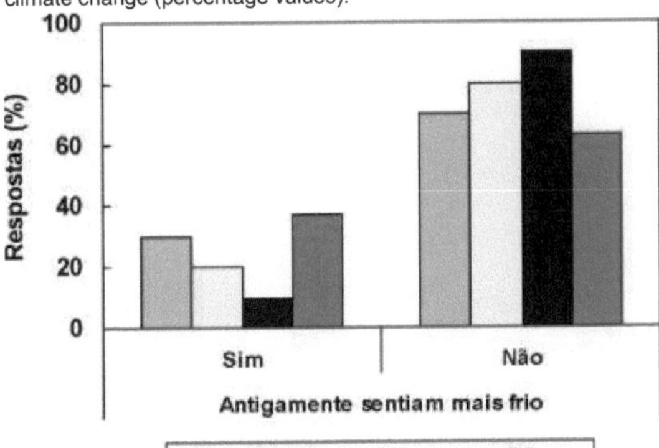

Fig. 22 - Sensitivity to climate change (percentage values).

As can be seen in figure 23, the majority of respondents say that the cold now lasts less time than it used to. In the case of respondents in the village of Lungongo, this percentage represents the totality of respondents.

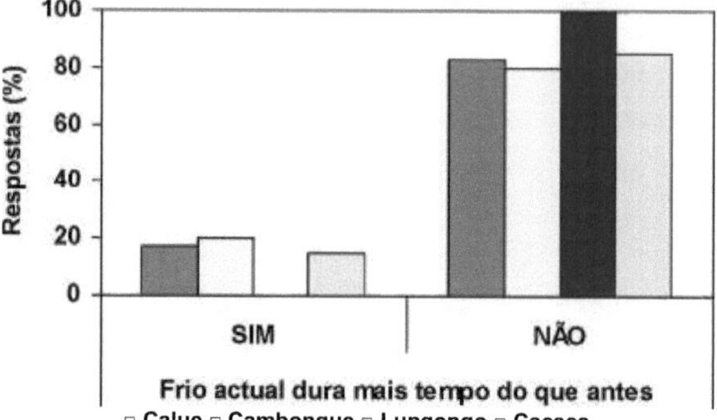

Fig. 23 - Sensitivity to climate change (percentage values)

They were also asked about the dust that is currently observed in dry weather. The data was very significant, especially for the village of Cacaca where 100 per cent of respondents said that there was more dust nowadays (figure 24).

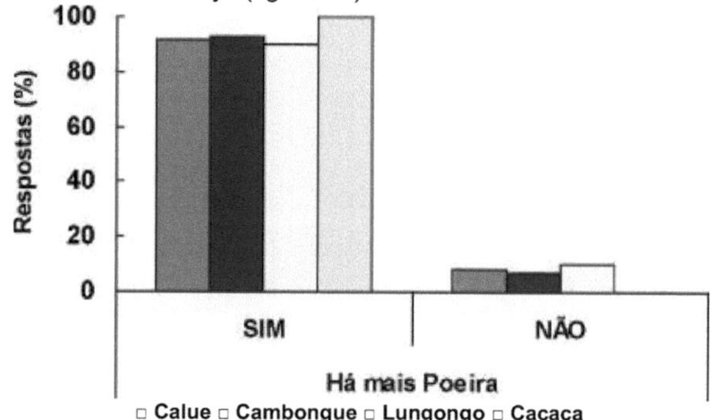

Fig. 24 – Sensibilidade à poeira devida à retirada da cobertura do solo e das cortinas quebravento (Valores percentuais).

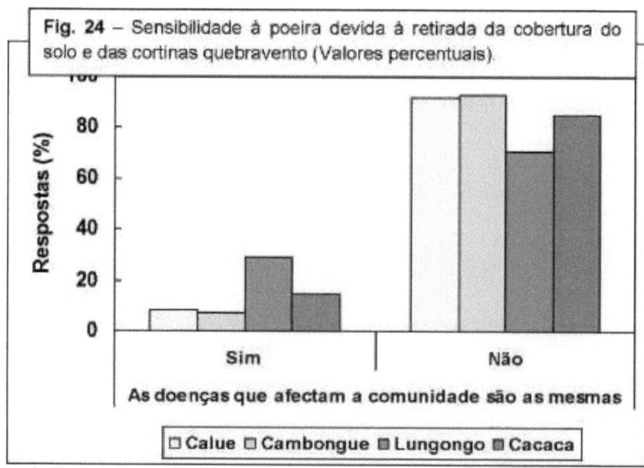

Fig. 25 - Changes in the pattern of illness due to temperature, dust and the quality of the water they use for drinking and cooking (percentage values).

As a result of changes in temperature, soil cover and the quality of the water they use for drinking and cooking, the pattern of illnesses, according to the respondents, has also changed. The value of the answers given should be confirmed with other, more in-depth actions.

3.7. Soil analyses

Table 14 - Soil analysis results

Location samples	pH H2O	pH KCl	Materia Organics (%)
Otchumbo 1	**4,5**	**4,4**	**1**
Otchumbo 2	**4,6**	**4,4**	**0,9**
Ombanda	**4,3**	**4,2**	**0,76**
Ongongo	**4,3**	**4,1**	**1,6**
Onaka	**4,2**	**4,1**	**1,2**

Studies carried out by Sanchez et al. (1984) reveal that over the last 30 years, soil fertility and yield levels have declined in large parts of tropical Africa because nutrient loss through leaching, erosion and harvesting has not been compensated for by natural processes and fertilisation.

The analyses carried out are not sufficient to make a complete diagnosis of the state of those soils. However, they were as good as they could be and give clear indications of the chemical and even physical poverty of the soils in question. The pH values in water are practically the same as those obtained in potassium chloride, which leads us to assume that the capacity to retain and absorb nutrients such as calcium and magnesium is almost zero. The levels of aluminium, manganese and zinc are probably close to toxic, creating an inhospitable environment for cultivation. With the existing pH values, phosphorus is completely inaccessible to plants. Most of these soils have extremely low levels of organic matter, so urgent measures need to be taken to improve their fertility. The use of agroforestry systems could be one way of doing this. The organic matter content should be improved, as far as possible, to no less than 2% (*CIAT*. 2006).

A good level of organic matter in the soil is important for crops and fulfils the following functions:

- It favours soil structure, leading to the formation of more stable aggregates that facilitate good water and air circulation in the soil, as well as root penetration, and reduce the risk of erosion;

- It increases the soil's water retention capacity, making it less sensitive to drying out, which is particularly important in light-textured soils;
- It is a source of nitrogen, sulphur and other nutrients for plants and improves the soil's ability to retain these elements;
- It increases the ability of plants to fix certain toxic elements, which are then absorbed in smaller quantities;
- It supports the soil's biological activity, which is ensured by fauna and a large number of microorganisms that make the soil a living environment;
- It contributes to the fixation of carbon dioxide (CO_2), reducing its concentration in the atmosphere.

One of the ways to achieve this is by periodically incorporating organic correctives.

Farmers' knowledge is not a panacea, but their knowledge and opinions on factors such as soil types, nutritional content, composting and crop response to organic and inorganic amendments directly affect their decisions. Our research approach uses dialogue between farmers to create "dynamic knowledge" to solve soil fertility problems. Understanding the processes that lead to the evolution of this "dynamic knowledge" is as important as the knowledge itself.

Analyses of the data lead us to table 15, which contains suggestions for the use of Agroforestry Systems. The table below describes some suggestions for the use of Agro-Forestry Systems according to the existing problem (Medrado, 2000).

Table 15 - Suggestions for using Agroforestry Systems.

PROBLEM	ALTERNATIVE
Lack of firewood	Living fences, plots of trees in pastures, woody trees in kitchen gardens and as shade for coffee, cocoa and other crops,
Erosion	Tree corridors on contour lines and appropriate trees along water sources
Degraded soils	Plant trees or shrubs that fix nitrogen and have deep roots between annual crops, use of legumes in the soil.
Strong winds	Plant trees or shrubs that fix nitrogen and have deep roots between annual crops, use of legumes in the soil.
Lack of food and shade for animals	Fodder trees or shrubs in living fences
Human food (Quantity and Diversity)	Vegetables, fruit and animal husbandry
Land demarcation	Living fences
Stabilising migratory agriculture and the reduction of socio-economic risks in the production system	Diversity in production with trees, animal crops, wood, medicinal plants and vegetable gardens
Excessive competition for water, light and nutrients between trees and associated crops	Modify existing systems with drastic pruning, elimination of certain trees. Modify the components of the lower level.

CONCLUSIONS

Farmers in the villages in the study area, in general, do not use good agricultural practices (contour lines, crop rotation and composting), but they do adopt some traditional agricultural practices (vipangas-camalhoes) which can perform functions similar to those of contour lines. They also burn and cut down trees and these practices have contributed to the process of soil degradation.

The main types of ploughs found in these villages are: Otchumbo, Ongongo, Ombanda and Onaka, most of the peasants have them in combinations (Otchumbo+Ongongo+Onka, Otchumbo+Ongongo, Otchumbo+Onaka).

The methods used to maintain the fertility of the soils found are the fallow method, which is most practised in the villages of Cacaca and Lungongo because there is still forest to be cleared. The image of soil degradation seen and felt by the community itself has led them to accept Agro-Forestry Systems as a way of making a significant contribution to changing this situation.

According to the evaluation carried out in these villages (Calue and Cambongue), the Agro-Forestry Systems that are suitable for the production units of peasant families are the Alley cropping system, which associates rows of trees with strips of annual crops that may or may not be shade-tolerant; Taungya system - This system helps to contain the advance of gullies, reduce strong winds and consequently dust, while at the same time restoring soil fertility. Species such as eucalyptus can be used for this.

RECOMMENDATIONS

- The Faculty of Agricultural Sciences, together with local NGOs, should organise training courses on good agricultural practices for these families.
- Civil society organisations in partnership with the Caala municipal administration should encourage the creation of community nurseries and ponds to mitigate the dust issue, reduce wind problems, develop reforestation programmes and introduce agro-forestry systems to restore some balance to the ecosystem.
- Members of the affected community must stop all activities that could trigger desertification, soil erosion or degradation and loss of land fertility (exploitation of inert materials such as stones, sand, avoidance of burning crop fields and wasteland).

BIBLIOGRAPHY

1. ALMEIDA, D. G. 2001. The construction of agroforestry systems based on local ecological knowledge (The case of family farmers working with agroforestry in Pernambuco), pp 238 f.
2. ALTIERI, Miguel A. 2002. Agroecology: a scientific basis for sustainable agriculture. Guafoa; Agropecuari,. 592p.
3. ASSAD. M.L.L. & ALMEIDA. J. 2003. Agriculture and Sustainability Context, Challenges and Scenarios. In: Ciencia e ambiente/ Universidade Federal de Santa Maria. UFSM, p. 5-14.
4. ASSIS. R. L. 2003. Agroecological Practices in Family Production in the Centre-South of Paraná. In: Ciencia e ambiente/ Universidade Federal de Santa Maria. UFSM,p.61 - 72.
5. BEER, J.; LUCAS, C; KAPP, G. 1994. Reforestation with permanent agrosilvicultural systems versus pure plantations Agroforesteria en las Americas,.v. 1, n. 3, p. 2125.
6. BOLFE, E. L. SIQUEIRA. E. R & BOLFE. A.P.F. 2003. Successional Agroforestry Systems: An Agroecological Practice. In: Ciencia e ambiente/ Universidade Federal de Santa Maria. UFSM,.p.86 - 94.
7. BUZA, Alfredo Gabriel. 2006. Potentialities and Socio-Economic Perspectives of Agroforestry Systems in the Municipality of Buco Zau, Province of Cabinda, Republic of Angola Belem, EMBRAP
8. BRIENZA J, S. 1986. EMBRAPA CPATU/PNPF agroforestry programme for the Brazilian Amazon. Documentos EMBRAPA/CPATU, n. 9, 11p.
9. CAMPELO, E.F.C.; SILVA, G.T.A.; NOBREGA, P.O.; VIEIRA, A.L.M.; FRANCO, A.A & RESENDE, A.S. 2006. Implementation and management of SAFs in the Atlantic Rainforest The Embrapa Agrobiologia experience. In: Agroforestry systems: scientific bases for sustainable development. Campos dos Goytazes, RJ: Universidade Estadual do Norte Fluminense Darcy Ribeiro, p. 33-42.
10. CHRISTINA H. 2002. Introduction and adoption of new production systems. Kenya.
11. CIAT. 2006. Creating dynamic knowledge for integrated soil fertility management in western Kenya, Africa.
12. CARVALHO, J. E. U. 2006. Utilisation of fruit species in agroforestry systems in Amazonia. In: GAMA-RODRIGUES, A. C. et al. (Ed.) Agroforestry systems: scientific bases for sustainable development. Campos dos Goytacazes, RJ: Universidade Estadual do Norte Fluminense, p. 169-176.
13. DUBOIS, J.C. L. 1989. Agroforestry: an alternative for sustainable rural development. Agroforestry Newsletter, REBRAF, VOL 1, N. 4, P. 1-7.
14. FAVARETO. A & BRANCHER. P. 2005. Territorial development in the Ribeira Valley and Pronaf-infrastructure projects - analysis and recommendations Research Report of the "Proposed Studies for the Dynamisation of Territorial Economies Programme" promoted by the Secretariat for Territorial Development of the Ministry of Agrarian Development - 1st cycle Available at: http://serv-sdt- 1.mda.gov.br/gnc/gnc/ep/estudos/SP_ValeRibeira.doc.
15. GOLLEY, F.B. 1983. Ecodevelopment. In; F.B. (Ed) Tropical rain forest ecosystems structure and function. Amsterdam: Elsevier Scientific Publishing Company,p.
16. GOMES. J.C.C. & BORBA. M. 2003. Limits and Possibilities of Agroecology as a Basis for Sustainable Societies In: Ciencia e ambiente/ Universidade Federal de Santa Maria. UFSM,. p. 5-14.
17. GOTSCH, E. 1995. The rebirth of AS-PTA agriculture, Rio de Janeiro. 22p.
18. HERNANDEZ, S.; BENAVIDES, J. 1995. Potencial forrajero de especies lenosas de los bosques secundarios de El Peten, Guatemala Agroforesteria en las Americas,. v. 2, n. 6, p.15-22.

19. HOWARD, Sir Albert; JESUS, E. L. 2007. An agricultural testament Sao Paulo: Expressao Popular. 360p.
20. KHATOUNIAN, C. A. 2001. The ecological reconstruction of agriculture Botucatu: Agroecologica,. 348p.
21. LOEDEMAN, J.H. 2005.Topographical surveys applied to rural areas. Agromisa, Netherlands.n.6
22. MACHADO, R.C.R.; GAMA-RODRIGUES. F.E.; MOQO. K.S. & GAMA RODRIGUES. A.C. N. 2006. Biological Attributes in Soils under Cocoa Agroforestry Systems: A Case Study. In: Agroforestry systems: scientific bases for sustainable development. Campos dos Goytazes, RJ: Universidade Estadual do Norte Fluminense Darcy Ribeiro, p. 33-42.
23. MARTINEZ Higuera, H. A. 1989. El componente forestal en los sistemas de finca de pequenos agricultores. Turrialba: CATIE, 79 p. (CATIE. Boletin Tecnico, 19). Programme for Sustainable Agricultural Production and Development, Forestry and Agroforestry Production Area.
24. MAZOYER, Marcel. 1987. Synthesis report In: Colloquium Dynamics of Agricultural Systems. Paris: INRA.
25. MEDRADO, M. J. S. 2000. Agroforestry systems: basics and indications. In: Reforestation of rural properties for productive and environmental purposes: a guide for municipal and regional actions. Brasilia: Embrapa Communication for Technology Transfer; Colombo, PR: Embrapa Floresta, p. 269-312.
26. MIRANDA, E. M.; VALENTIM, J. V. 1998. Establishment and management of living fences with multi-purpose tree species. Rio Branco: EMBRAPA-CPAF, 4 p. (EMBRAPA-CPAF. Comunicado Tecnico, 85).
27. OLIVEIRA Neto. 2004. Initial viability of a taungya system with Eucalyptus pellita Mell. and Phaseolus vulgaris L. in Paty do Alferes, RJ. In: CONGRESSO BRASILEIRO DE SISTEMAS AGROFLORESTAIS, 5., Curitiba. SAFs: development with environmental protection: proceedings. Colombo, PR: Embrapa.
28. UN. 2006. Deforestation/forestation rate in the world. News Centre.
29. OTS/CATIE. 1986. Agroforestry systems: principles and applications in the tropics. San Jose: Organisation for Tropical Studies/CATIE, 818p.
30. OTAROLA, A. 1999. Black wood living fences: agroforestry practice for dry season sites marked Agroforestena en las Americas,. v. 2, n. 5, p. 24-30.
31. OVIEDO, F.; VALLEJO, M.; BENAVIDES, J. 1994. Modulos agroforestales para la produccion de leche con cabras Agroforesteria en las Americas,.v. 1, n. 2, p.23-27.
32. PACHECO, Fernando. 2005. Catena utilisation scheme in Huambo in: CESD. Angola. Catholic University.
33. PECK, R. B.; BISHOP, J. P. 1992. Management of secondary tree species in agroforestry systems to improve production sustainability Amazonian Ecuador. Agroforestry Systems,.V. 17, p. 53-63.
34. PENEIREIRO. F.M. 1999. Agroforestry systems driven by natural succession: a case study. Postgraduate course in Sciences (Master's dissertation). Piracicaba, Escola Superior de Agronomia Luis de Queiros, , 138p.
35. PENEIREIRO. F.M. 2006. Fundamentals of Successional Agroforestry Seminar on Agroecology and Sustainable Rural Development. Centre for Agricultural Sciences - UFSC - Florianopolis,. p. 96 - 103.
36. PEREIRA. J.P.; LEAL. A.C. & RAMOS. A.L.M. 2006. Agroforestry systems with rubber trees. In: Agroforestry systems: scientific bases for sustainable development. Campos dos Goytazes, RJ: Universidade Estadual do Norte Fluminense Darcy Ribeiro, p. 33-42.
37. RIBEIRO, M. B; FREITAS. C. S. 2007. Women's work in agroforestry systems in Alto Jequitinhonha; Available at: < http://www.aba- agroecologia.org.b>. Accessed on: 02 June

2010, 12:34:55

38. SANTOS, A.C. 2005. The contradictions of the market economy: a look at income from agroecological agriculture. Agriculturas, v. 2, n° 3, p.07-11, oct.

39. SANCHEZ, J.H. 1984. Continuous Cropping potential in the Upper Amazon Basin.University Presses of Florida.

40. SOMARRIBA, E. 1995. Guayabo in pastures: establishment of live fences and recovery of degraded pastures. Agroforestena en las americas,.v 2, n. 6, p. 2729.

41. VAREJAO, P. 2009. Incaper encourages the implementation of agroforestry systems in the north of the state Available at: http://www.es.gov.br/site/noticias. Accessed: 26 June 2010, 20:23:44

42. VIVAN, J.L. 1998. Agriculture and forests: principles of a vital interaction. Guafoa: Agropecuaria, S.l

43. WEBER, F. R. and STONEY C. 1986. Types of simultaneous agro-forestry systems

44. YOUNG, A. 1991. Agroforestry for soil conservation CAB International, 275p. (ICRAF Science and Practice of Agroforestry, n.4).

ANNEXES

Enquiry.1

NAME
AGE-
COMMUNITY

Do you have a plough?	S	N	
1. MINING SITE	LAVAGE AREA If you use a plough, how many days does it take to work your plough?		
	A-ONAKA		
B-OMBANDA	B-OMBANDA		
C-ONGONGO	C-ONGONGO		
D-OTCHUMBO	D-OTCHUMBO		
2. This ploughing you have, the soil is still there. good?	This good	This more or less	This evil
3. How many years can a plough last produce well without resting?	4 a 6	6 a 10	More
4. For the land to recover: a) Fallow the land (leave the plot fallow) earth to rest for a while)	YES	NO	
b) Cut down a new area in the forest to have more fertile land?	YES	NO	
5. If you do fallow land, how many years does it take for the land to recover? Tell us about it	5-10	More than 10	
5. Do you practice Good Agricultural Practices? Tell us about it	Composting? Crop rotation? Level curves		
6. If you have a guarantee of recovering land that no longer yields much, isn't it better to stay in the same place than look for another site?	YES	NO	
7. Are you willing to try Agroforestry Systems on your land? Tell us about it	SIMNAO		
8. Which species do you want to introduce?	Forest species		
	Forest species + fruit trees		
	Forest species + fruit trees + improvement species		
. 9) Which of your plots are you willing to experiment with agroforestry systems? Area			

A-ONAKA		
B- OMBANDA		
C- ONGONGO		
D- OTCHUMB O		

. 10. Do you think it could be an experience on community land rather than individual land? Tell us about it	YES	NO	
. These areas are yours and no one will take you in. . 11. if the systems are good, will it encourage other people to join?	YES	NO	
. 12) What are the expected advantages of introducing SAFs (Agroforestry Systems)? Crop diversification More products for sale Income generation Maintaining soil fertility	YES	NO	
Pest and disease control Explain a little			
. 13) What are the main concerns about joining SAFs? Explain a little.	TECHNICIANS FINANCIAL FEAR FAIL		
. 14. what questions do you want to ask?			
NAME- AGE- COMMUNITY-			

WATER		
15. Does the rain come at the same time as before?	YES	NO
16. Does it rain well?	YES	NO
17. Is there enough water for agriculture?	YES	NO
18. When the crop is still in the field.	YES	NO
19. Does it sometimes rain a lot?	YES	NO
20. Does it rain little in others?	YES	NO
a). Do you sometimes have to water the maize in the rainy season?	YES	NO
21. In the old days, did you have to water too, or did the rainwater arrive and distribute it well until it was time to harvest?	YES	NO
22. Do you only grow crops in the rainy season?	YES	NO
23. Do you also grow in dry weather?	YES	NO

24. If so, what crops do they grow?	
25. What is the watering method? Sprinkling Drop by drop Bucket	
26. Do rivers carry water like they used to?	YES NO
27. Do the rivers have clean water?	YES NO
Do the rivers have rubbish in them? 29. Who causes this rubbish? Wind	
People	
30. What rubbish is in the river water?	
. 31. How long does the water in the cacimbas last?(Indicate month) 32. Is that enough to withstand the dry weather?	
SIMNAO	
. 33. Where does the water you drink and use for cooking come from?	Rio Crank Cacimba
. 34. Is the water good or does it cause diarrhoea? 35. Do you boil the water?	YES NO YES NO
NAME- AGE- COMMUNITY-	

FORESTS	
36. What is the state of the forest in your area?	Good More orBad Less
37. If it's bad, why?	Are they cutting down too many trees? SIMNAO Are they burning too many trees? YES NO
38. Which trees are near the community?	Which trees are cut down the most?

Eucalyptus	
Pinheiro	
Cedar	
Onduko	
Omanda	
Other	

Eucalyptus	
Pinheiro	
Cedar	
Onduko	
Omanda	
Other	

39. What benefits do you get from the forest?	
40. Do forest products help to reduce hunger and poverty? Explain a little	
41. What do you do with the eucalyptus you cut down?	
42. What do they do with the omanda and onduko they cut?	
43. If they cut it down, don't you think they should plant new ones?	YES NO
44. Did anyone tell you that you should grow saplings to replace the trees you cut down?	YES NO
45. Does anyone in this community grow eucalyptus or pine trees?	YES NO
. 46 Where did the seedlings come from?	EDA ADRA NGO IDF
. 47. Does anyone in this community have a nursery for onduko or omanda or do you think it's not necessary for these species?	YES NO
. 48. Is the forest important for the temperature? 49. Is the forest important for rain?	YES NO YES NO
. 50. Do you usually burn in your community?	YES NO
. 51. Is it as cold as it used to be this year?	YES NO
. 52. Were you colder in the past? . 53 Did the cold last longer in the past?	YES NO YES NO
. 54 Is there a lot of dust? . 55. Are the diseases that children get now the same as in the past?	YES NO YES NO

SUGGESTIONS

Enquiry.2

ONDUKO-
ALIMA-
YMBO

. OKUETE EPYA?	Otcho	Atchoko

. EPYA OKUETE	
B-OMBANDA	
C-ONGONGO	
D-OTCHUMBO	

Nda okulima lo osalua, oloneke vingami?

A-ONAKA	
B-OMBANDA	
C-ONGONGO	
D-OTCHUMBO	

. Epya okuete osi likuete ongusso?	Likuet	inse	kalikuete

. Alima angami epya kalipuyka masi	4 to 66 to 10+ of 10
likala longuso lyo kuima longuso?	-
. Otcho Epya likale longuso:	Otcho Atchok
epya yene lisokupuyka?	
Osendasenda usengue otcho	Otcho Atchoko
osange epya longuso?	
. Nda epye lisokupuyka, alima angami	5-10+ out of 10
ipuyka? Vangula pokamue	
Okuete ovitua viwa vyokulima? Vangule pokamue . Nda epya lia kuata vali onguso, opongolola vali epya?	Ombolela? Osole oku soveka akunla? osole okulinga Capenga tchasokelela ndovava atai? Otcho Atchoko
. Oyongola oku seteka akunla le pako ale loviti kepya limosi? Vangula pokamue	Otcho Atchoko
. Oviti oyongola oku seteka vyane?	Oviti yovusengue Oviti yovusengue kuenda le pako - Oviti yovusengue, epako kuenda evi vietcha onguso vepya -
. Kepya lyove pi oyongola oku seteka akunla le pako loviti kepya limosi? Otchikap	

A-ONAKA		
B- OMBANDA		
C- ONGONGO		
D- OTCHUMBO		

. Oyongola oku seteka Kepya lyove ale vepya youini? Vangula pokamue	Otcho		Atchoko
. Epya tua popya ndoto yove. . Nda oseteko yatcho ey ya tcha tchiwa olonguisako vakuavo?	Otcho		Atchoko

. Nye okasi okutalamela lo seteko ey (SAFs)? Okutenga akunla Unja yokulandisa Otchitumbulikila yofeto Okukapa onguso kepya Okutata ovipuka kuenda ovivey kakunla Vangula pokamue		
. N tchikumbatisa koseteko ey (SAFs)? Vangula pokamue.	Oseteko	
	Olombongo	
	Usumba yoko ponya	
. Oyongola okupula tchimue?		
ONDUKO- ALIMA- YMBO-		
OVAVA		
. Ombela isole okufetika okuloka vosay imosi?	Otcho	Atchoko
. Ombela iloka tchiwa? Ovava atenlisa akunla eto?	Otcho	Atchoko
	Otcho	Atchoko
. Nda akunla akasi vepya. Kuli olondjandja akuti ombela iloka tchalua? Kuli olondjandja akuti ombela iloka tchitito?	Otcho	Atchoko
	Otcho	Atchoko
. a). Kotembo yombela osole oku Yulisa epya ye epungo? Ulima Wapita wa regaleli epungo ale ovava ombela atenlisile akunla?	Otcho	Atchoko
	Otcho	Atchoko
. Kondombo lika osole okulima?	Otcho	Atchoko
lima kokuenye? Nda otcho nye olimalima?		
. Nda mupi osole oku regala? Ovava okunyava ethosi la ethosi otchukuata kokuyengeka		
. Volui andi muli ovava?		

. Ovava volui alyela?	
. Ovava volui a sila? Elye ofunguisa ovava olui?	
Ofela	
Omar Elino	io Iyovava olui pi yatunda?
. Onjombo itakaka osay) Ytenlisa okuenye?	osay ipi?(Tukula
Otcho	Atchoko
. Ovava okunyua lo kuteleka pi isole oku tundila?	volui Crank
	Vonjombo
. Ovava anena ovipulukalo? Osole okufelula ovava?	Otcho — Atchoko Otcho — Atchoko
ONDUKO- ALIMA- YMBO-	
USENGUE	
. Nda mupi ikasi usengue?	Tchiwa — Inse tchasepul;
. Nda tchasepula, momo nye?	Vakasi okuteta oviti? Vakasi oku temia oviti? Otcho — Atchoko Otcho — Atchoko
. Oviti vipi tukuete otchipepi lyovimbo?	Oviti vipi vakasi oku temia?

Eucalyptus	
Pinheiro	
Cedar	
Onduko	
Omanda	
Other	

Eucalyptus	
Pinheiro	
Cedar	
Onduko	
Omanda	

	Other	
. Osilivilo ypi ikuete usengue?		
. Apako yovisengue itepulula esako kuenda ondjala? Vangula pokamue		
. Nye osole okulinga localipto utete?		
. Nye osole okulinga lomanda kuenda londuko uteta?		
. Oviti vinuay?	Otcho	Atchoko
	Otcho	Atchoko
Nda osole okuteta oviti, kutuete ekulihinso yoku kunla vikuavo?		
. Kavako sapuile ulono wo kukunla oviti yovisengue kuna osole okuteta?	Otcho	Atchoko
	Otcho	Atchoko
. Vimbo kamuli omuno osole okukunla oviti vyo calipi ale yo pinho?		
. Pi ytunda oviti osole okukunla?	EDA	
	ADRA	
	NGO	
	IDF	
. Mulo vimbo kuli omuno osole okulinga embumba yo onduko ale omanda ale katchitava okuvikunla?	Otcho	Atchoko
. Usengue ikuatisa utanya?	Otcho	Atchoko
Usengue ikuatisa okuloka ombela?	Otcho	Atchoko
. Vimbo vasole okutemia usengue?	Otcho	Atchoko
. Ombambi elimenli lya livokya?	Otcho	Atchoko
. Ulima wa pita ombambi lya pyanlile?	Otcho	Atchoko
. Ulima wa pita ombambi lya livokya?	Otcho	Atchoko
. Kuli oneketela?	Otcho	Atchoko
. Uvey ikuata omala kotembo ey imosi ayo volotembo vapita?	Otcho	Atchoko

ESOKOLUILO

ANNEX.2

Table I - Absolute answers given by respondents on the subject of rainfall

	It rains at the same time		It rains well		Enough for agriculture		Little rain	
	Yes	No	Yes	No	Yes	No	Yes	No
Calue	0	60	0	60	17	43	60	0
Cambongue	0	15	0	15	7	8	15	0
Lungongo	0	30	0	30	12	18	0	30
Cacaca	0	20	0	26	5	21	26	0

Table II - Data on river flow and river water quality.

	Taking them with old	rivers n water o amente	Clean water		There's rubbish	
	Yes	No	Yes	No	Yes	No
Calue	0	59	25	34	36	22
Cambongue	2	13	5	10	10	5
Lungongo	0	30	30	0	0	30
Cacaca	0	26	24	2	2	24

Table III - River water quality data.

	Who causes this rubbish				Where they draw water for drinking and cooking					
	People		Winds		Cacimba		Rivers		Withstands dry weather	
	S	N	S	N	S	N	S	N	S	N
Calue	28	0	25	0	60	0	0	0	28	30
Cambongue	10	1	10	0	10	0	0	0	6	9
Lungongo	0	0	0	0	26	0	7	0	9	19
Cacaca	0	0	0	0	26	0	0	0	13	13

Table IV - Availability of water from the cacimba during dry weather

	July		August		September		October	
	Yes	No	Yes	No	Yes	No	Yes	No
Calue	20	0	0	0	18	0	38	0
Cambongue	0	0	0	0	0	0	12	0
Lungongo	0	0	0	0	0	0	0	13
Cacaca	0	0	0	0	0	0	0	14

Table V - Water quality data

	Aguapara drinking causes diarrhoea		You've been boiling the water	
	Yes	No	Yes	No
Calue	45	15	52	8
Cambongue	9	6	12	3
Lungongo	19	11	21	8
Cacaca	15	11	19	6

DEDICATION

I offer this work to my two princesses: Laureta *Graqa* Miguel and *Gersine* Miguel, who have provided synergies for the transformation of my life, giving me the strength to fight, dream and win.

ACKNOWLEDGEMENTS

To the sovereign and triune God for life, health and every blessing.

To my wife Laureta Gra$a and my daughter Gersine Miguel, for their affection, encouragement and help at all times and under all circumstances.

To my parents, Anacleto Ramalho and Laureta Usanga, for their principles, love and crucial support at every moment of my life.

To my brother Agostinho (St Augustine) for his tireless companionship and help at every moment of my life.

My sisters: Celina, Joana, Edviges for their affection.

To Professor Virg^ia Quartin, my supervisor, for her trust, valuable teachings, attention and time throughout the construction of this thesis and throughout my academic career.

To Professor Valente, for his valuable contributions and wise life guidance, in the belief that every journey begins with the first step and with rigour.

To the teachers on the course: Dr Ginhas Alexandre, Dr Orlis Barbara, Professor Belarmina, Dr Carlos da ConceiQao, Dr Rodrigues, Professor Daisy Maria Delgado, Dr Raymundo Tielvis, Professor Rolando Ramirez, Professor Reynaldo, Professor Maribel Mesa, Professor Maria do Ceu Babo Cordeiro for the knowledge they passed on.

To my fellow students, especially Alister Pinto, Alfredo Ahave, Antonino Camutali, Jorge Gomes, Felizardo Capepula, Adilson Rodrigues, Feliciano Mingondji, for their companionship.

The Angolan ADRA, and in particular the director of the Huambo antenna, for their valuable support and for all the facilities granted for carrying out the surveys as part of a project by this Angolan non-governmental organisation.

To Mr Helio Tiago and Mr Carlos Chivembe for their help and support at all times.

To everyone named or unnamed, my eternal gratitude.

SUMMARY

The main aim of this work was to make a diagnosis of the agro-environmental situation in the villages of Calue, Cambongue, Lungongo and Cacaca, belonging to the municipality of Caala, and to analyse the possibility of these farmers adopting and adapting Agro-Forestry Systems (AFS) as a way of recovering degraded areas. Agricultural practices were analysed in the different types of ploughlands, i.e.: onaka (low ploughlands); ombanda (ploughlands located on the edges of the onaka); ongongos (high ploughlands) and in the otchumbo (backyard) and the impact of these practices on natural resources. The methodology used was surveys and semi-structured conversations. A total of 131 surveys were carried out (72 women and 59 men). The villages of Calue and Cambongue were our main object of study because they presented the worst picture in terms of deforestation, poor soil cover, the presence of ravines, and urgently needed a recovery plan. Most of the farmers surveyed in these villages, despite having different types of plough, choose to cultivate in the otchumbos, on the one hand because they are closer and make cultivation easier, and on the other because they benefit from domestic waste and the soils are more fertile. The soil analyses carried out show that the pH varies between 4.2-4.6 and the organic matter content between 0.7-1.6%. SAFs, in which trees or shrubs are used in conjunction with agricultural crops or animals in the same area, are a form of sustainable management of agricultural systems.
Keywords: *Surveys, semi-structured conversations, degraded areas, agro-forestry systems*

Jose Eduardo dos Santos University
Faculty of Agricultural Sciences
Huambo

Department of Forestry Engineering.
Final course work for a degree in Agricultural Engineering.
Agricultural Engineering

Title: Diagnosis of the natural resources of the villages of Calue and Cambongue, with a view to indicating the Agro-Forestry Systems for the recovery of degraded areas

Author: Agnelo Chiunde Miguel
Supervisor: ***Professor Dr Virginia Maria Lacerda Quartin***
Nº Registration: 0109/2012
Huambo, July 2013

Jose Eduardo dos Santos University
Faculty of Agricultural Sciences
Huambo

Department of Forestry Engineering.
Final course work for a degree in Agricultural Engineering.
Agricultural Engineering

Title: Diagnosis of the natural resources of the villages of Calue and Cambongue, with a view to indicating Agro-Forestry Systems for the recovery of degraded areas
Author: Agnelo Chiunde Miguel
Supervisor: *Professor Dr Virginia Maria Lacerda Quartin*
Nº Registration: ***0109/2012***
Huambo, July 2013

Printed by Books on Demand GmbH, Norderstedt / Germany